**Protecting children's health
in a changing environment**

The World Health Organization was established in 1948 as the specialized agency of the United Nations serving as the directing and coordinating authority for international health matters and public health. One of WHO's constitutional functions is to provide objective and reliable information and advice in the field of human health. It fulfils this responsibility in part through its publications programmes, seeking to help countries make policies that benefit public health and address their most pressing public health concerns.

The WHO Regional Office for Europe is one of six regional offices throughout the world, each with its own programme geared to the particular health problems of the countries it serves. The European Region embraces some 880 million people living in an area stretching from the Arctic Ocean in the north and the Mediterranean Sea in the south and from the Atlantic Ocean in the west to the Pacific Ocean in the east. The European programme of WHO supports all countries in the Region in developing and sustaining their own health policies, systems and programmes; preventing and overcoming threats to health; preparing for future health chal¬lenges; and advocating and implementing public health activities.

To ensure the widest possible availability of authoritative information and guidance on health matters, WHO secures broad international distribution of its publications and encourages their translation and adaptation. By helping to promote and protect health and prevent and control disease, WHO's books contribute to achieving the Organization's principal objective – the attainment by all people of the highest possible level of health.

Protecting children's health in a changing environment

Report of the Fifth Ministerial Conference on Environment and Health

WHO Library Cataloguing in Publication Data

Protecting children's health in a changing environment : report of the Fifth Ministerial Conference on Environment and Health.

1. Child welfare 2. Climate change 3. Environmental health – trends 4. Health policy 5. Health promotion 6. Congresses 7. Europe

ISBN 978 92 890 1419 9 (print) (NLM Classification: WA 30)
ISBN 978 92 890 1420 5 (ebook)

ISBN 978 92 890 1419 9

Address requests about publications of the WHO Regional Office for Europe to:

Publications
WHO Regional Office for Europe
Scherfigsvej 8
DK-2100 Copenhagen Ø, Denmark

Alternatively, complete an online request form for documentation, health information, or for permission to quote or translate, on the Regional Office web site (http://www.euro.who.int/pubrequest).

© **World Health Organization 2010**

All rights reserved. The Regional Office for Europe of the World Health Organization welcomes requests for permission to reproduce or translate its publications, in part or in full.

The designations employed and the presentation of the material in this publication do not imply the expression of any opinion whatsoever on the part of the World Health Organization concerning the legal status of any country, territory, city or area or of its authorities, or concerning the delimitation of its frontiers or boundaries. Dotted lines on maps represent approximate border lines for which there may not yet be full agreement.

The mention of specific companies or of certain manufacturers' products does not imply that they are endorsed or recommended by the World Health Organization in preference to others of a similar nature that are not mentioned. Errors and omissions excepted, the names of proprietary products are distinguished by initial capital letters.

All reasonable precautions have been taken by the World Health Organization to verify the information contained in this publication. However, the published material is being distributed without warranty of any kind, either express or implied. The responsibility for the interpretation and use of the material lies with the reader. In no event shall the World Health Organization be liable for damages arising from its use. The views expressed by authors, editors, or expert groups do not necessarily represent the decisions or the stated policy of the World Health Organization.

Contents

Abbreviations vi

Introduction 1

1. Progress in environment and health, 1989–2010 5

2. Environment and health challenges in a globalized world: role of socioeconomic
   and gender inequalities 9

3. Implementing CEHAPE 12

4. Investing in environment and health 17

5. Dealing with climate change in Europe: challenges and synergies 21

6. Future of the European environment and health process 26

References 29

Annex 1. Parma Declaration on Environment and Health and Commitment to Act 32

Annex 2. The European environment and health process (2010–2016): institutional framework 38

Annex 3. Parma Youth Declaration 2010 41

Annex 4. Declaration of the European Commission 44

Annex 5. Programme 45

Annex 6. Core publications 49

Annex 7. Pre-Conference and side events 50

Annex 8. Participants 58

Abbreviations

CEHAP	children's environment and health action plan
CEHAPE	Children's Environment and Health Action Plan for Europe
CO_2	carbon dioxide
DPSEEA	Drivers – Pressures – State – Exposure – Effects – Actions (model)
EC	European Commission
ECDC	European Centre for Disease Prevention and Control
EEA	European Environment Agency
EFSA	European Food Safety Authority
ENHIS	European Environment and Health Information System (of the WHO Regional Office for Europe)
EU	European Union
HEAT	health economic assessment tool
IGOs	intergovernmental organizations
NEHAP	national environment and health action plan
NGOs	nongovernmental organizations
ODA	official development assistance
OECD	Organisation for Economic Co-operation and Development
PM_{10}	particulate matter less than 10 μm in diameter
RIVM	National Institute for Public Health and the Environment (the Netherlands)
RPGs	Regional Priority Goals (of the CEHAPE)
SAICM	Strategic Approach to International Chemicals Management
SARS	severe acute respiratory syndrome
THE PEP	Transport, Health and Environment Pan-European Programme
UNDP	United Nations Development Programme
UNECE	United Nations Economic Commission for Europe
UNEP	United Nations Environment Programme
UNFCCC	United Nations Framework Convention on Climate Change
WHY	World Health Youth (Communication Network on Environment and Health)

Introduction

The series of WHO ministerial conferences on environment and health is unique in bringing together different sectors to shape European policies and actions on the environment and health. The first four conferences were held in Frankfurt, Germany in 1989, Helsinki, Finland in 1994, London, United Kingdom in 1999 and Budapest, Hungary in 2004 *(1–4)*. Focusing on the measures that countries could take to protect children's health from environmental risk factors, the Fourth Ministerial Conference adopted the Children's Environment and Health Action Plan for Europe (CEHAPE) *(5)*. An intergovernmental mid-term review, held in 2007 in Vienna, Austria *(6)*, noted the progress made in acting on the Budapest commitments and identified the priorities for the Fifth Ministerial Conference.

A range of environmental risk factors threatens health: inadequate water and sanitation, unsafe home and recreational environments, lack of spatial planning for physical activity, indoor and outdoor air pollution, and hazardous chemicals. Recent developments – including financial constraints, broader socioeconomic and gender inequalities and more frequent extreme climate events – amplify these threats. They pose new challenges for health systems and environmental services to improve health through effective environmental health interventions, as well as to safeguard the environment.

The Fifth Ministerial Conference on Environment and Health was therefore convened in Parma, Italy on 10–12 March 2010, to enable ministers of health and of the environment, key partners and experts to assess the progress made since the first conference. Organized by the WHO Regional Office for Europe and hosted by the Government of Italy, the Conference offered governments an opportunity to renew the pledges made in 2004 and to address new challenges and developments. Notably, the Fifth Ministerial Conference took place in an era in which governments faced new global challenges to improving both health systems' performance and collaboration between the health and environment sectors to ensure better environments for health. The Conference also marked the latest milestone in the environment and health process in the WHO European Region, which Member States had initiated over 20 years previously.

The Conference was the product of extensive consultation with representatives of Member States, international organizations, the research community and civil society. WHO held high-level, Region-wide intergovernmental preparatory meetings in Germany, Italy, Luxembourg, Spain and other Member States; subregional meetings for south-eastern Europe and the newly independent states; and meetings of many technical working groups.

The Conference agenda encompassed several main priority areas. First, participants:

- assessed the progress made in environment and health in Europe since the first European conference in 1989, and the current environment and health situation in the European Region, focusing particularly on the countries of south-eastern and eastern Europe, the Caucasus and central Asia;

- evaluated the impact of the environment and health process in Europe; and

- reviewed the extent to which decisions taken at previous conferences had been implemented and where further action was needed.

The Regional Director addresses a packed audience © WHO/Andreas Alfredsson

Then they reviewed measures that could be taken to address socioeconomic, gender, age and other inequalities in environment and health. Third, the participants addressed an area of increasing concern: the effects of climate change on health and the environment. Finally, they discussed how to move forward in the environment and health process in Europe, particularly how to strengthen local and subregional implementation.

The major policy outcome of the Conference was the Parma Declaration (Annex 1); other outcomes comprise annexes 2–4. The Declaration outlines the actions that ministers agreed to take on the priority issues addressed in the Conference programme (Annex 5), in collaboration with the European Commission, international and intergovernmental organizations (IGOs), civil society and other partners. Annexes 6–8 list the various working documents, policy briefs and background documents that informed the discussions; related events taking place before and during the Conference; and the participants, respectively.

Zsuzsanna Jakab, WHO Regional Director for Europe, opened the Conference. Pietro Vignali, Mayor of Parma, and Vincenzo Bernazolli, President of the Province of Parma, welcomed the participants. Both emphasized the need to give effect to integrated, intersectoral policies and to reduce the environmental effects on health, particularly in the difficult current economic situation.

In her opening address, Stefania Prestigiacomo, Minister of Environment, Land and Sea of Italy, confirmed that better health is the objective of all environmental policies. Protecting children's health in a changing environment, the theme of the Fifth Ministerial Conference, is of particular importance because of children's greater vulnerability to environmental hazards and the worrying trends in their health status. Ferrucio Fazio, Minister of Health of Italy, noted that environmental factors account for over 30% of diseases in children aged under 5 years. In Italy, close cooperation between the environment and health ministries resulted in the adoption of a national health care plan in 2008 that draws attention to, for example, the health effects of chemical pollutants and calls for preventive action by not only the health sector but also such sectors as environment and transport.

Zsuzsanna Jakab acknowledged the support received from Member States for the WHO European Centre on Environment and Health, with its offices in Rome and Bonn, and previously in Bilthoven; that had significantly increased the WHO Regional Office for Europe's capacity to provide countries with top-level technical advice. Much was achieved during the 20 years of the European environment and health process, but the burden of disease from environmental determinants of health in the WHO European Region remains substantial. More powerful and more comprehensive policy responses are needed to ensure that diseases are prevented and health outcomes further improved. One major cause for concern is the continued growth of inequalities in exposure to environmental risks. A study launched by WHO to coincide with the opening of the Conference (7) reveals that the social distribution of environmental exposures and related deaths and disease shows very significant inequalities both between and within countries.

These disconcerting trends and statistics form a very strong argument for a renewed strategic alliance between the environment and health sectors. If the right preventive policies are adopted and applied, the overall burden of disease can be reduced by almost 20%, while well-tested environment and health interventions could save 1.8 million lives a year in the WHO European Region. To achieve this, the consideration of health and health inequities should be mainstreamed into all public policies and national development programmes, particularly those in the transport and industry sectors. Equally, simultaneous work at the international, national and local levels could maximize the impact of joined-up policies. Only through a proactive and inclusive process of policy development and advocacy can other parts of government and society be convinced that health is not only a public expenditure but also a resource for a better economy, better quality of life and ultimately a more just and equitable society.

WHO needs a new vision for European health policy and a new, comprehensive and value-based strategy that makes health a horizontal government responsibility. That means continuing to collaborate closely and engaging in a deeper dialogue with key partners such as the United Nations Economic Commission for Europe (UNECE), United Nations Environment Programme (UNEP) and other United Nations bodies, as well as the Council of Europe, the World Bank and the Organisation for Economic Co-operation and Development (OECD).

After acknowledging the important role played by the European Environment and Health Committee, under its joint chairpersons Corrado Clini and Jon Hilmar Iversen, in following up the outcomes of previous ministerial conferences and planning the current one, Zsuzsanna Jakab paid tribute to Dr Jo E. Asvall, who had served as WHO Regional Director for Europe for 15 years and, sadly, passed away in February 2010. In his last speech to staff at the Regional Office, 12 days before his death, he had urged them to be courageous and willing to take risks; Ms Jakab emphasized that only by working together and taking risks would the Conference participants be able to translate the values of human rights, universality, solidarity, equity, participation and access to quality health care into tangible health benefits in societies.

Ján Kubiš, UNECE Executive Secretary, said that he believed that the European environment and health process is unique since it rightly puts the two sectors on an equal footing. They are the driving forces behind efforts to secure human health and, in a wider sense, behind sustainable development. Two unique instruments gave the clearest evidence of the success of the collaboration of UNECE and the WHO Regional Office for Europe: the Transport, Health and Environment Pan-European Programme (THE PEP) (8), and the Protocol on Water and Health to the 1992 Convention on the Protection and Use of Transboundary Watercourses and International Lakes (9). Nevertheless, other legal instruments also link environment and health, such as the UNECE Protocol on Strategic Environmental Assessment (10) and the Convention on Long-range Transboundary Air Pollution (11). Promising areas for further collaboration include a possible framework convention on affordable, healthy and green housing, as well as the third round of environmental performance reviews conducted in the countries in transition in the region covered by UNECE. The Seventh Ministerial Conference of the "Environment for Europe" process will be held in Astana, Kazakhstan in 2011.

Margaret Chan, WHO Director-General, addressed participants by video link, since she was visiting Bangladesh and the Maldives to see the effects of climate change on them at first hand. Recalling the start of the European environment and health process at the first ministerial conference in Frankfurt, she commended the governments of countries in the Region on being among the first to focus on environmental factors as the primary causes of multiple widespread health problems, and to see them as an opportunity for population-wide prevention, and especially as a resource for the promotion of healthy lifestyles. The conferences have given the Region a head start in tackling issues that are now of concern in every part of the world.

During the Fifth Conference, participants would look in particular at the role played by social and gender inequalities in the distribution of environmental hazards, and the environmental problems and needs in the newly independent states and countries of south-eastern Europe. Dr Chan warmly supported efforts to give people living in those countries a level of protection that matches the standards in place elsewhere in the Region.

Lastly, the Conference was held at a time when many countries were seeking ways to put the findings of the Commission on Social Determinants of Health into practice in a whole-government approach to health *(12)*. That means addressing the root causes of ill health as far upstream and as comprehensively as possible. One of the biggest challenges is to persuade other government sectors to include health concerns in their policies; the European environment and health conferences offer a model of collaboration in that area as well. They have given a straightforward message: multisectoral cooperation for better health is indeed feasible.

1. Progress in environment and health, 1989–2010

Regional and global assessment

Information collected through the WHO Regional Office for Europe's European Environment and Health Information System (ENHIS) *(13)* and two surveys enabled an assessment of the major trends in progress towards achieving the four Regional Priority Goals (RPGs) of the CEHAPE: clean water, injuries and physical activity, clean air, and reduced environmental hazards such as chemicals and noise.

Overall environment and health conditions in the WHO European Region are better than in 1989, when the first ministerial conference took place, but further improvement is still possible.

- Thousands of cases of diseases related to drinking-water are registered every year, even in developed countries, and many more go undetected.

- Access to safe water has grown in most countries; in 10 Member States in the Region, however, over half the population in rural areas still has no access to safe water.

- Road traffic injuries have fallen by a third since the early 1990s.

- One year of life expectancy is lost due to air pollution in many areas of Europe. Levels of particulate matter less than 10 μm in diameter (PM_{10}) have remained unchanged for 10 years, but could be cut by 50% if all currently feasible measures were implemented. Indoor air pollution is still poorly addressed.

- The risk of asthma is 50% higher for people living in damp and mouldy dwellings, and over 20% of households in many countries report problems with dampness.

- As to chemicals, some positive effects of intervention are observed, such as a drop in dioxin levels in breast-milk.

- One in five people is exposed to noise at night at levels high enough to disturb sleep and raise levels of cardiovascular risk.

Responses to a survey on environment and health policy in 40 countries confirmed that the health and environment sectors often work together to develop and implement policies involving the agriculture, education and transport sectors. Most encouragingly, the environment and health process is moving from reactive preventive measures to the proactive creation of better environments.

At the global level, 25% of disease is estimated to be associated with environmental risk factors. The climate change debate has created an opportunity, as many parties are eager to reach an agreement after the somewhat disappointing outcome of the 2009 United Nations Climate Change Conference *(14)*. A more flexible approach is most likely to succeed. A strategic alliance between the environment and health sectors is essential, as the

A lively panel discussion © WHO/Andreas Alfredsson

two sectors pursue the same ends; for example, most of the actions that reduce levels of carbon dioxide (CO_2) emissions benefit health.

The environmental health agenda needs to be revitalized through more primary prevention. By widening its scope, it can include not only water and sanitation, indoor and outdoor air and the reduction of toxic substances but also work through healthy cities and urban planning, occupational health and reduced exposure in the home. In addition to improving the environment, action in all these areas will also help to reduce noncommunicable diseases and prevent communicable diseases. For example, primary prevention measures in traffic have multiple positive effects on health: reducing obesity, injuries and depression, increasing social capital and cutting cardiovascular diseases.

The move to a greener economy, while a necessity for economic growth, also brings health benefits. A focus on higher-quality food and more efficient waste disposal, for instance, helps to mainstream health in other areas. Health is an added value that policy-makers in all sectors should use as a driving force. Further, the health sector needs to lead by example, by cutting its own CO_2 emissions. Greening the health sector is possible in both developed and developing countries.

The European Region has achieved a great deal in the last 20 years, and the world is counting on its leadership and experience to pave the way forward.

Useful tools: a legal instrument, a programme and joint work

Ten years old, the Water and Health Protocol addresses RPG1 of the CEHAPE: to achieve access to safe water and sanitation for everyone, with a particular focus on vulnerable groups *(5,9)*. The Protocol was needed because 13 000 children die every year from poor-quality drinking-water; 140 million people do not have a household connection to a drinking-water supply; 41 million people lack access to a safe drinking-water supply, and 85 million people do not have improved sanitation. Climate change and emerging trends, such as protozoan infestations of drinking-water supplies and the proliferation of *Legionella* spp., make the need more urgent.

The Protocol is a powerful tool because it is legally binding on its signatories, making its conditions hard to ignore even in times of financial crisis. It provides the institutional framework for adaptation to climate change, the integration of policies and the implementation of other conventions and conditions. It is also a concrete and practical tool, with achievable targets and a reporting mechanism to measure continuous progress that facilitates each country's compliance. By connecting water and health authorities, the Protocol obliges them to work together in a multisectoral fashion and at the international level.

Without safe water, there can be no health. The technical solutions are known; what is now needed is the political will. Countries should therefore ratify and implement the Protocol, use it to help fulfil their commitments – such as achieving the Millennium Development Goals *(15)* – and European Union (EU) directives, to reduce health inequalities related to socioeconomic factors, gender and age, and to ensure adequate resources for implementation (see Annex 7).

THE PEP was launched in 2002 *(8)* as a result of the 1989 European Charter on Transport, Environment and Health *(16)*. Countries are encouraged to join it for various reasons, the most compelling being that it contributes to not only economic growth but also improved health and environment. THE PEP has four main priorities: integrating environment and health into transport policies, shifting transport demand to sustainable mobility, improving urban transport and facilitating the consideration of cross-cutting issues in specific areas.

Its main achievement is the development of tools and methods such as the health economic assessment tool (HEAT), which allows the economic valuation of transport-related health effects; the toolbox for policy-makers; the clearing-house for exchanging knowledge and information; and guidance for integration of environment and health concerns into transport policy. THE PEP has seen a shift in thinking, encouraging more environmentally friendly and healthy forms of urban transport, and raising awareness of cross-cutting issues in countries in the eastern part of the European Region. Countries that have benefited financially through more efficient implementation of THE PEP tools and methods include Austria, the Czech Republic, Hungary and Sweden. Countries have shared the national plans they develop, strengthening partnerships as cooperation evolves. The goals for 2009–2014 are securing sustainable and efficient transport systems, reducing transport-related emissions and shifting to safe and healthy modes of transport.

Similarly, the work done jointly by WHO, UNECE and UNEP shows that the intersectoral approach is the way forward: representatives from environment and health organizations increasingly attend each other's meetings, creating a synergy in which health is often the common element. In February 2010, for instance, conferences of the Parties to the Basel, Rotterdam and Stockholm conventions – which address hazardous waste, pesticides and industrial chemicals, and persistent organic pollutants, respectively – held simultaneous meetings for the first time.

Recognizing that the legally binding nature of such instruments strengthens implementation, Member States are considering tackling the effects of mercury on health and the environment through a new agreement. Article 24 of the Convention on the Rights of the Child *(17)* relates to health, making it one of the most important conventions underpinning the ideals of the CEHAPE. The Strategic Approach to International Chemicals Management (SAICM), a policy framework to foster the sound management of chemicals, is a strong multistakeholder strategy of the United Nations, WHO and business interests, with health at the core *(18)*.

The links between health and the environment are easy to see. For example, contaminated water can kill and low-quality water can make people ill, so the proper management of water ecosystems is vital. The threat of climate change makes action to protect the evenironment and health even more imperative.

Achievements of the environment and health process

A panel of experts who have participated in it discussed the achievements of the environment and health process. It has raised environment and health higher on the political agenda in the WHO European Region, and catalysed change in other regions, too. It had resulted in the creation of a powerful tool within the WHO Regional Office for Europe: the WHO European Centre for Environment and Health, which can assess changes and propose new policies. The lessons learnt in Hungary serve as an example of progress in countries; the process has enabled the country to build its own policies, based on the convergence of environment and health challenges.

Further, the process has broken down the barriers not only between environment and health but also between government and nongovernmental entities, nongovernmental organizations (NGOs) and IGOs, and professionals and non-professionals. The links between the environment and health had joined health promotion as twin concerns of the WHO Regional Office for Europe. Nevertheless, panellists questioned whether existing intersectoral collaboration was sufficient, and whether WHO could practise stronger advocacy in the style of NGOs.

The panellists cited the Conference itself as proof that the process is working, but called for the expansion of ownership of the process to include all sectors and wider society. To come on board, these new partners need to feel a personal sense of responsibility for the process. The advocacy required to bring in other sectors, such as transport, should appeal to the emotions, as well as provide information. One of the factors in CEHAPE's success in attracting partners and resources for implementation is the emotional element of its focus on children, as well as its originality in involving the young.

Despite the successes discussed, sectors still have separate agendas, and arguments need to be tailored to each to bring them on board. Health is a persuasive argument, however, especially when combined with financial ones. Environment and health must not be seen as costs but as investments. In fact, the environment can be seen as a booming sector, investing in a sustainable future. Some existing instruments requiring intersectoral collaboration may need upgrading to become more effective. Making them legally binding would ensure that ministers do not renege on them during hard times.

Governments need proof to show the difference that the environment and health process can make, but they still lack such information. As journalists are keen to take part in advocacy, they should be given the information they need to play their part.

The environment and health process needs to be more systematically expanded to other sectors, through the approach of including health in all policies. Further, technical experts need to consider that the environment and health process operates on a longer term than the political cycle, and provide politicians with arguments that they can use. For the future, the process needs to be more flexible, promoting intersectorality; focus more on implementation; address climate issues; and continue involving young people to ensure sustainability.

2. Environment and health challenges in a globalized world: role of socioeconomic and gender inequalities

Equity in health, climate and the environment

Two recent publications assess environmental inequalities and health in Europe and the United Kingdom: *Closing the gap in a generation*, the final report of the Commission on Social Determinants of Health chaired by Sir Michael Marmot, and *Fair society, healthy lives. Strategic review of health inequalities in England post-2010* (the Marmot review) *(19,20)*. According to the latter, life expectancy in England and Wales had increased in 1972–2005, but continued to be lower for unskilled than skilled workers. Data on the impact of the social gradient in England showed that, for each year, if everyone had the mortality of those with a university education, 202 000 people aged 30 years or more would not die prematurely, thereby gaining 2.5 million years of life.

A conceptual framework to reduce health inequities and improve health and well-being for all should rest on the creation of an enabling society that maximizes individual and community potential and ensures that social justice, health and sustainability are at the heart of policies. To do this, three key actions are recommended.

- Policies and interventions that both reduce health inequalities and mitigate climate change should be prioritized.

- Planning, transport, housing, environmental and health policies should be integrated.

- Locally developed and evidence-based community regeneration programmes should be supported, especially those that remove barriers to community participation and action and that emphasize a reduction in social isolation.

The challenges of equity in health, the steps made towards it and its relationship with climate and the environment can be seen from four angles: equity; gender; climate, environment and health; and the social determinants of health inequalities. The struggle against poverty demands moral, political and social development. Human health is an all-purpose goal and an essential requirement for individual freedom: where equity in health exists, everyone benefits. The WHO Constitution *(21)* states: "the enjoyment of the highest attainable standard of health is one of the fundamental rights of every human being without distinction of race, religion, political belief, economic or social condition". Unfortunately, this definition did not take account of gender, and gender differences in health risks due to environmental exposures persist. Progress towards gender equity is being made, but is unequal.

The situation of the interrelated areas of climate, environment and health is increasingly disquieting, jeopardizing the quality of natural and vital resources and endangering human existence. Development based on ecological and social ethics therefore needs to be promoted.

Policies linked to energy, agriculture and the exploitation of the earth must not disregard health and social analyses that take account of underprivileged populations. Evidence of gains in health conditions and policies is extensive in the EU, where most countries experience the advantages of a social model of health. The past few years have seen an increased commitment to direct tackling of the social determinants of health and the resulting inequalities. The promotion of health in all policies contributes to population-wide risk prevention, with the most visible effects among underprivileged populations.

Despite these advances, fairness in the health sector must be further promoted to prevent the growth of inequities. Further, several EU resolutions surprisingly couch some exhortations to health in economic terms. People's health has its own intrinsic value, beyond its importance to the economy.

Some challenging inequalities

A panel of country representatives provided examples.

In Germany, data support the finding that social status affects health and longevity. The financial crisis is exacerbating social inequalities in risk, especially among children and elderly people. Though limited data are available, these effects need analysis. Minority and migrant populations are also at increased risk. An improvement in primary health care and public health is needed, with a focus on nutrition and general public awareness of health. Germany is working to identify and reduce these environmental injustices and plans to foster greater cooperation and focus on this subject.

In Malta, the increasing numbers of illegal migrants coming from sub-Saharan and northern Africa over the last decade are a cause for concern. On their boat journey to Malta, these people suffer many health risks, including exposure to the weather, overcrowding and even drowning, as well as dehydration, minor burns, scabies and respiratory and gastrointestinal illnesses. After arrival, the migrant population suffers the additional threats of exposure to local pathogens, occupational health and safety problems and sexually transmitted infections, along with the risk of mental ill health due to feelings of isolation, and the traumas faced in their countries of origin or on the journey. In Malta, migrants also concentrate in particular areas, increasing the population density and thus the pressure on the local infrastructure, particularly sewage and waste.

The populations of Malta and the European Region as a whole are entitled to the same environmental conditions and health care, and migrant populations should be a particular focus owing to the risks they face and their generally poorer living conditions. As climate change increasingly threatens Africa, the likelihood of climate refugees rises, with subsequent effects on Malta, particularly the availability of food and water. Malta is seeking comprehensive solutions and making increased efforts to return illegal migrants to their countries of origin, while supporting measures to encourage legal migration.

The Russian Federation supports the need to strengthen the systematic monitoring of the health and environmental situation and to use these data to tailor specific programmes to address them. The exposure of pregnant women to chemicals, at work or in the general environment, is of great concern in the country, due to the possible effects of these exposures on the fetuses. In addition, decreasing chemical exposure during the first year of life is very important to prevent adverse effects on children's development and health. WHO has a clear role to assist here, with its enormous capacity to disseminate evidence and strengthen health professionals' capacity.

Chemical safety is also a concern in Slovenia, as people in all countries have the right to live and work in safe environments. Inequalities both within and between countries therefore make it essential for all countries in the European Region to cooperate. Chemicals and chemical safety are key areas where inequality is clearly evident, yet the lack of reliable data and biomonitoring related to health and the environment hinders progress. The legacy of obsolete pesticides, along with chemical contamination from industrial activities, is an additional burden. Slovenia has developed a strategy to strengthen the engagement of the health sector in SAICM and is focusing on improving the management of obsolete pesticides and other chemicals, a topic for discussion at the Sixty-third World Health Assembly and in the EU. The health sector needs to engage to a greater degree with the SAICM initiative, as this sector deals with the consequences of chemicals management.

In 2010, Slovenia is hosting the first meeting of a working group to prepare a strategy for strengthening the health sector's engagement in chemical management. Joint action of the health and other sectors, along with closer cooperation between Member States and international organizations, can reduce the differences between countries, thus protecting the most vulnerable populations and ensuring a safer environment today and for future generations.

Several participants contributed to the discussion, agreeing that environmental policies need to focus more on population health. Belgium supports the use of norms and standards, but promotes the need for criteria for access to environmental health services and the need for locally based policies and pledges in which the health sector and environment sector work together. To support this cross-sectoral approach, Belgium plans to promote the deeper incorporation of social determinants of health in environmental health policy. In Portugal, health equity is a main component of the national health plan for 2011–2016, which supports citizen empowerment to stop social exclusion and promotes early access to daycare, particularly among migrants.

To support the initiatives on socioeconomic and gender inequities, areas where health policy should tackle health inequity include education, health and environment, and the inclusion of health in all policies. Other areas in which countries can learn from each other include the development of standards and preventive programmes. WHO has an important role in bringing countries together and identifying what areas would benefit from such convergence. Finally, it is important to remember that a growing economy is not always related to improving health, and some efforts to maintain economic strengths harm the health of poor communities.

In summary, the following issues are key.

- Although intersectoral collaboration is difficult and challenging, it is feasible and a key component in the inclusion of health in all policies, which WHO will continue to support. Countries need to share their experiences, however; for example, Portugal has actively pursued the intersectoral and health-in-all-policies approaches, with resulting improvements in life expectancy and the health of the population.

- The economic crisis can be viewed as a new opportunity to adjust priorities to invest more in health promotion and disease prevention, and to include environmental health in the broader concept of public health.

- Action on the social determinants of health needs to be promoted and this requires leadership and information, which the WHO Regional Office for Europe can provide.

3. Implementing CEHAPE

CEHAPE awards

NGOs organize the competition for the CEHAPE awards to highlight and reward good practice in children's environment and health. The Health and Environment Alliance and the Eco-Forum presented the second CEHAPE awards to eight inspiring and innovative projects that have made a major contribution to improving children's environmental health (22). These projects are run by youth associations, women's organizations, schools, institutes and other NGOs.

Inspired by the launch of the CEHAPE (5) and the Declaration of the Fourth Ministerial Conference on Environment and Health (23) in 2004, and first presented during the WHO intergovernmental mid-term review hosted by Austria in 2007 (6), the awards are intended to emphasize that local action is crucial where children play and live. Prizes were awarded in eight categories, four relating to the RPGs, two to growing challenges and two to potential solutions (Table 1). The 20 judges awarded marks to 114 projects, submitted from 31 countries over 3 months. The projects show concrete benefits, a partnership approach, originality, transferability, cost–effectiveness and ability to raise awareness. Each of the eight winners was presented with a cheque for €1000 by a panel of seven representatives of health and environment ministries and one from the European Commission (EC).

Table 1. CEHAPE awards

Category and topic	Country with winning project
RPGs	
Water and sanitation	Lithuania
Accident prevention and physical activity	United Kingdom
Air quality	Belgium
Hazardous chemicals and radiation	Russian Federation
Challenges	
Mobility	Austria
Climate protection	Armenia
Solutions	
Youth participation	Russian Federation
Schools	Tajikistan

- In Lithuania, schoolchildren collected water samples from rural wells for analysis and gave feedback to their communities, resulting in the improvement of the quality of the water.

- In the United Kingdom, Child Safety Week provided millions of parents with safety information, such as keeping matches and cleaning products away from children and practising road safety with them, through easy-to-use materials and ideas for events.

Young recipient of a CEHAPE award

- In Belgium, primary schools improved their indoor air quality. Children's awareness of indoor air quality was raised through games, songs and the use of a child-friendly CO_2 monitor that turns red when the air is poor.

- In a mining area in the Russian Federation, the top layer of soil in kindergartens was cleaned, reducing by 50% the number of children with blood lead levels above the safety threshold. Although expensive, the project was effective and is being copied in Kazakhstan.

- A campaign by students in Austria raised awareness of poor public transport alternatives to the private car, and increased by 50% the number of pupils and teachers who bicycle to school.

- In Armenia, a solar energy panel installed by a women's group at a kindergarten created a warmer and cleaner indoor environment for the children, saved money by cutting energy bills and reduced CO_2 emissions.

- In the Russian Federation, a youth group began an interactive environmental education programme, to share experts' knowledge with 12 young trainers and, through them, with hundreds of students. This motivated them to embark on a range of activities, such as collecting waste, recycling paper, adopting healthier lifestyles and raising awareness.

- In Tajikistan, students developed a PC-based manual to promote activities to make their schools more environmentally friendly: cleaning them up, recycling waste, distributing clean water, providing low-cost heating and making posters. The incidence of diarrhoeal diseases has dropped and recycling covers the costs of the actions.

Lessons to be shared

The panel of ministers and the EU representative shared their experiences, in answer to questions from youth representatives and the two NGOs.

In Azerbaijan, access to clean water and sanitation remains a huge challenge in rural areas. Securing funds from the EU is a high priority to ensure that mobile water purification plants can continue to increase thousands of villagers' access to clean water. Austria prioritizes healthy transport as a means of meeting its climate goals, and the panel member invited the youth representative from Bosnia and Herzegovina to Austria to share its experience of free transport for young people. In Belgium, "green ambulances" diagnose indoor air quality, and financial incentives and product norms are used to improve the quality of building materials. A chemicals action plan is being developed in Denmark. It applies the precautionary principle to the possible risks from exposure to a combination of chemicals in daily life, and vulnerable groups such as pregnant women and mothers are informed about chemicals in everyday products.

Young people's participation is a natural corollary to their being the targets of many health initiatives, such as those on nutrition, mental health, alcohol and tobacco. The EU led the way with a young people's conference in 2009, which produced a road map for youth health. Norway's environment and health strategy for children and young people (2007–2016) promotes active youth involvement, and a new planning and building act developed by the health and environment ministries requires children to have good environments in which to grow, and local governments to ensure that children and young people can actively participate in planning.

In Armenia, incorporating environment and health issues into the school curriculum requires a major shift in attitude among teachers, as well as in supporting legislation. Nevertheless, schools should encourage pupils' interest in the topic by raising their skills, motivating them to act and supplying them with examples of good practice and the necessary books and information. Funding is a major limiting factor. In the Netherlands, moves to mitigate or adapt to climate change in the area of clean transport and better indoor environments in schools are recognized as also benefiting health.

Asked what actions were needed to strengthen the practices cited, most of the panel agreed that legislation is important. It should be used to ensure that young people are involved in planning. Existing legislation should be used, rather than more enacted. Countries should share their experiences, especially as this reinforced the value of a bottom-up approach. Social partnerships, such as NGOs at both the national and local levels, are also key. Bold action can be taken: where lead is banned to protect children from exposure, for example, substitutes have been found as a result. Implementation is important: good ideas need to be put into practice and the commitment of civil society and young people is essential. EU support, particularly in the form of common legislation, is essential, while communication, education and empowerment are also core ingredients.

Achievements, challenges and a possible way forward

The Fourth Ministerial Conference on Environment and Health, in 2004, redefined the environment and health challenge by reinforcing the enduring importance of environmental health concerns and by extending the reach and relevance of environment and health activity to align with a new and challenging agenda for public health and health improvement *(23)*. This so-called ecological public health now has renewed importance.

Countries have taken great strides since 2004. A questionnaire on the CEHAPE was sent to the 53 Member States in the WHO European Region, and 46 responded. The key findings are as follows.

- Out of 53 countries, 49 now have environmental health focal points.

- While 30 have children's environment and health action plans (CEHAPs), 12 are developing them and 4 have not begun. Some CEHAPs are related to national environment and health action plans (NEHAPs); others are linked with action plans related to children, and 12 are stand alone.

- The CEHAPE has positively influenced intersectoral collaboration, public information and awareness, interventions to improve children's health and environment, the development of monitoring and information systems, and the development of national CEHAPs.

- The challenges countries face include: insufficient capacity and resources and consequently unsustainable actions, insufficient intersectoral collaboration, the low relative importance of environment and health in national policy-making, and a lack of methods to engender cross-cutting work, to identify evidence-based interventions and to link to policy.

- The next steps are: raising the profile of environment and health at the national level and finding effective ways to engage policy-makers and politicians, linking children's environmental health to other complex policy agendas, obtaining WHO support for national actions, sharing insights and experiences about conceptual and methodological challenges, and developing tools that can be readily adapted to different national contexts.

In this era of ecological public health, all the determinants of health and well-being are important. A new way of presenting the problem conceptually might be a modified DPSEEA (Drivers – Pressures – State – Exposure – Effects – Actions) model *(24)*, where a context (social, cultural, demographic, economic, behavioural) section is incorporated into the exposure and effect components and implemented in practice through: framing the problem, quantifying the pathways, performing a gap analysis (research, policy, and effectiveness) and building systems to advise policy-makers on appropriate actions. Complexity must be embraced.

Benefits of CEHAPs

A national CEHAP was initiated in Austria in 2005. Through the cooperation of the ministries of health and the environment, a national coordinator and task force were established. Other stakeholders were engaged in the process, particularly young people, along with representatives of other sectors: social welfare, economics and finance, energy, transport and education. Pilot projects began in 2005 and an awareness campaign was launched. Strong political willingness and a clear strategy resulted in the commitment of human, technical and financial resources.

The experience suggests that national CEHAPs are to be recommended and that, although cooperation can be very fruitful, it needs to be supported by a supranational initiative to strengthen pan-European cooperation with:

- joint projects and partnerships;

- capacity building and support;

- liaison with THE PEP;

- consideration of emerging issues such as climate change and nano technology;

- a target-oriented approach; and

- an upgrading of the CEHAPE.

Several participants shared their experiences of implementing CEHAPs. The heart of France's environmental health plan is its CEHAP, the achievements of which include increased access to kindergartens, decreases in noise and improvements in air quality. In Belgium, the plan for 2009–2013 is to establish priorities focusing on children and including human biomonitoring of exposure to heavy metals and chlorates, and research into asthma. The CEHAP is an essential tool and should be disseminated as a global approach with child-centred projects.

Montenegro has taken great strides since 2004, when it performed an environmental health performance review and developed a CEHAP through cross-sectoral collaboration. Environmental legislation has been aligned with that of the EU, but implementing and enforcing this new legislation will require the development of capacity, and the identification of funds and time. Malta included child-specific action in its NEHAP activities for 2006–2010. Young and intersectoral stakeholders are engaged in the process and a high-level environmental health committee meets regularly.

Challenges of implementing CEHAPs

A panel including country representatives and regional and youth representatives reviewed the challenges of implementing CEHAPs. In Cyprus, the main enabling factors are seen as promoting the initiative within government and establishing strong political will. EU policies provide an enabling framework, and the integration of policies and strategies to move towards a more holistic approach addresses some financial instruments as well. A defined budget allocation provides certainty of action. In the Republic of Moldova, the integration of health and environment into other sectors' policies and strategies is seen as beneficial. An evidence base is needed to support policy development.

The main challenge to implementing the CEHAPs in Poland and Portugal was knowing the priorities and formally appointing a CEHAP committee. In Poland, WHO's involvement in identifying priorities is an asset, as is the long-established collaboration between the health and environment sectors.

On the regional level, eight countries have cooperated on a programme on indoor air quality in schools, in which local and national efforts are required and flexibility is essential. The views of young people are best elicited through youth organizations, peer-to-peer activities, national committees, and studies and action on such issues as nutrition and smoking.

All panellists agreed that financing activities is a key challenge.

Participants made several proposals on how to move forward. A database of more and less successful examples of CEHAP activities should be developed, hosted by the existing systems of ENHIS *(13)* or the National Institute for Public Health and the Environment (RIVM) in the Netherlands. It should focus on the configuration of the family, placing children's needs first and engaging parents. Special standards for children should be developed. Child safety action plans are being developed in 25 Member States, using proven measures. Changes of government or government members can hinder progress and dissipate momentum. Evidence-based interventions should be used. Local-level governance has an important role and children's exposure to second-hand tobacco smoke is a concern.

In summary, action in countries needs:

- to increase the focus on children;
- to include scientific experts in the legislative process;
- to develop child injury prevention plans;
- to involve government agencies from the beginning in any interventions;
- to harmonize EU and national legislation;
- to use existing infrastructure for collaboration;
- to assess the comparative cost–effectiveness of synergy with a policy on climate change;
- to increasingly use and report to existing systems such as ENHIS; and
- to collect better data on the RPGs.

To enable the process and develop the current agenda, action is needed that is supported by strong political engagement, addresses challenges posed by climate change and nano technology, considers that all health determinants matter and includes the sharing of information and experiences.

In conclusion, work for child environmental health is essential at all levels. WHO has an important role in providing continued support. NGOs and other agencies need to participate to support the lobbying of leaders. From an ethical point of view, the people affected by inequalities, who are the most vulnerable in the current economic crisis, must be considered.

4. Investing in environment and health

Working with partners and stakeholders

City perspective

Pietro Vignali, Mayor of Parma, described the city's success in developing integrated policies to solve common problems, using the example of transport and mobility. Stimulated by a grassroots movement to reduce PM pollution, the city integrated its environmental policies with those of other sectors, such as infrastructure, health, mobility and transport, and introduced incentives to adopt good practices.

As a result, 90 km of bicycle lanes have been constructed, electric bicycles have been introduced, and Parma has moved from seventeenth to second place in a ranking of cities in environmental terms.

Subnational perspective

In a region of Sweden, a classic top-down approach led to some sophisticated epidemiological investigations, but they were considered useless for local authorities. Instead, a common aspiration to sustainable development in the region was agreed with municipalities and used as a tool for developing a public health policy. Proximity to local actors and the public enabled a constructive dialogue built on a certain degree of trust. On that basis, the considerable amount of information required was brought together, covering not only how health and health determinants are distributed in the population but also why they are distributed in that way and what kind of decisions is needed to reduce health and environmental inequities.

In addition to a formal structure at the local and regional levels, regions can benefit from membership of WHO's Regions for Health Network, a grouping that allows for systematic collaboration and exchange of experience *(25)*.

National perspective

At the national level, three main challenges in the multisectoral dimension of working with partners and stakeholders need to be faced: ensuring coherence of policy between various ministries, engaging different levels of government and involving NGOs. To ensure equal and well-functioning partnerships, it is important to adopt a common language, choose the right skill mix of collaborators, and respect and use existing structures whenever possible.

The area of diet and physical activity offers a good case study of the approach adopted in Switzerland. On the basis of a number of international instruments and policy documents – the 2004 World Health Assembly resolution on the Global Strategy on Diet, Physical Activity and Health *(26)*, the WHO European Charter on Counteracting Obesity *(27)* and the European Commission's white paper on a strategy for Europe on nutrition, overweight and obesity-related health issues *(28)* – a Swiss national programme on diet and physical activity was drawn up for 2008–2012. The programme was developed through a participatory process led by the Federal Office of Public Health and involving the Federal Office of Sports, Health Promotion Switzerland, representatives of the cantons and industry, and an alliance of NGOs and numerous other actors. All of these were also entrusted with implementing defined programme components. The programme uses a range of approaches: guarantees of food safety, economic support for voluntary measures and promotion of individual responsibility in a variety of target groups and settings. The Federal Office of Public Health operates a monitoring system on nutrition and physical activity, and promotes action in cooperation with private companies.

European perspective

While the founding regulation of the European Food Safety Authority (EFSA) emphasizes science-based policy and the separation of risk assessment from risk management, the resulting core value of independence does not imply isolation. On the contrary, one of EFSA's key roles is to coordinate networks of scientific excellence and stakeholders in the food chain.

EFSA increasingly needs to include environmental risk assessments in its work and to provide comprehensive responses using the full range of expertise at its disposal, so it cooperates with national food safety agencies, partner institutions of the EU and international counterparts. More than 350 scientific organizations lend experts each year to help EFSA build its risk assessment capacity. It maintains an important dialogue with the EC Directorate-General for Research and stakeholder organizations, through bodies such as a consultative group on emerging risks and a stakeholder consultative platform. In addition, it proposes to establish a standardized EU-wide food consumption database. EFSA's communication practices are regularly informed and updated by Eurobarometer surveys of risk perception among the public at large. These activities underscore the need to engage a wide range of actors in protecting public health.

From global to local perspectives

The European Environment Agency (EEA) focuses on the impact of environmental issues not only on Europe but also globally. Access to information and reporting is a challenging issue and data need to be timely, up to date and trustworthy. Current data flows show a cumbersome mechanism of data transfer through reporting. With the introduction of EEA's Shared Environmental Information System, electronic data input will provide a more rapidly available source of information, decrease costs and provide a more open form of information sharing, particularly as environmental issues cross borders. The recently launched Eye on Earth platform provides up-to-date information on air and water quality in Europe *(29)*. The system enables anyone to submit observations about perceived air or bathing water quality by SMS. Global Monitoring for Environment and Security provides *in situ* coordination services for land, climate and air monitoring, along with marine services and emergency response. A genuine opportunity exists for the environment and health community to reach out to a broader group of people through the greater use of and engagement in these services.

Needs for improved partnerships

In a panel discussion, panel members agreed that work with partners takes many forms. The EC has a tradition not just of consulting with partners but of establishing joint fora or platforms with them. The EU Platform for Action on Diet, Physical Activity and Health, for instance, is a well-structured mechanism for taking action on a set of joint commitments and monitoring implementation by means of common indicators *(30)*.

Intersectoral cooperation is perhaps more difficult to achieve in the public sector, although the emergence of civil society has led to a rapprochement of actors in that sphere. In the eastern part of the WHO European Region, however, countries have found it easier to initiate or maintain interministerial collaboration and harder to forge partnerships with civil-society organizations. The trade union movement offers governments a good route for reaching people at home through their work. A multisectoral approach should always include a youth element, to promote initiatives such as peer-to-peer education. Governments' role includes ensuring the framework within which stakeholders can become engaged.

More data and information are needed to gain a better understanding of stakeholders' perceptions in the area of risk assessment, for instance. Although many issues – such as the benefits of physical activity *(31)* or the adverse effects of night noise *(32)* – have already been thoroughly explored, more transparency and independent research could form the basis for greater public participation in risk management. Although a distinction must be made between science for research and science for decision-making, both are needed.

The importance of working in partnership with stakeholders is now widely recognized. It is time to look for action and results: strengthening networks of different partners, working with existing structures in the short term and making changes that will bear fruit in the medium and long terms.

Role of international financing mechanisms

Climate change is a great challenge and a threat to health, but also an opportunity. Structural transformation is needed to counteract this threat. The funds made available for such environmental changes create opportunities by financing the transition to better developmental paths, including environmentally friendly technologies and the creation of new jobs in a greener economy. Using funds to tackle climate change can benefit health at the same time. Thus, the links between health, the environment and young people can be tied to developmental finance.

The two main tools to tackle climate change are mitigation – reducing greenhouse-gas emissions – and adaptation through greener mechanisms. Yet their annual global costs are estimated at US$ 550 billion and US$ 86 billion, respectively, while official development assistance (ODA) for climate financing is limited to US$ 10 billion. Even the Copenhagen Green Climate Fund, which came out of the 2009 United Nations Climate Change Conference, has short-term pledges for only US$ 30 billion per year.

Private-sector financing must therefore be sought, but many countries, especially in the eastern part of the WHO European Region, need international support to secure it. Several barriers prevent access to these funds, such as countries' lack of both capacity and staff, the small amounts made available by the various funding bodies and limits on the countries that can benefit. Four main methods can bridge this funding gap:

- removing the barriers to implementation;

- scaling up existing financial mechanisms;

- bringing in new and innovative sources of finance; and

- building capacities to secure, absorb and deploy environmental financing.

The United Nations Development Programme (UNDP) is the biggest broker of finance to the environment, giving millions of dollars in direct grants and even more through co-financing, and ensures that most environmental projects include a health element. Nevertheless, UNDP's environmental finance services are already both complex and daunting and will become more so if new mechanisms are added. Those seeking funding therefore need to work more intelligently. For every US$ 1 the United Nations commits, US$ 45 can be raised from private sources.

Uzbekistan is an example of how funds can be leveraged for sustainable development. It is a carbon finance leader, with the most foreign investment in emission-reduction projects in the region, and among the 20 largest worldwide. UNDP invested US$ 260 000 in Uzbekistan and a green investment scheme was set up to reinvest proceeds from the sale of carbon credits into social, environmental and development projects. Thus, an initially modest outlay gave the country access to more funds from other sources. UNDP is eager to work with WHO to facilitate such schemes and build countries' capacity to access needed funds.

Experience with financing mechanisms

A panel described members' experiences with financing mechanisms. For example, Albania is implementing several projects in line with the CEHAPE and its NEHAP, mainly in the area of water and air quality, with support from both WHO and UNDP, and with funding from Austria, Italy and Germany. It has had to facilitate the legal basis for receiving funding from abroad, and now wishes to evaluate the direct effects of the projects on people's health.

Tajikistan uses several sources to finance environmental health projects proposed by both the environment and health ministries. These projects have had positive effects on, for example, the quality of and access to drinking-water and the incidence of water-related communicable diseases. The country has received funds from several international organizations and finance institutions through the Global Environmental Fund, to address air pollution, agricultural pesticides and climate change. Its main concern is to tackle lack of coordination and duplication of programmes.

The European Centre for Disease Prevention and Control (ECDC) funds research and projects on communicable diseases and climate change in EU countries and, through WHO, in the rest of the European Region; it stresses sustainable funding, not just short-term support. Such investments are well spent, as the financial consequences of neglecting such threats as severe acute respiratory syndrome (SARS) or foot and mouth disease far outweigh the costs of dealing with them. Constant surveillance is essential to monitor communicable diseases and the movements of vectors that result in the new geographical spread of the diseases they carry.

OECD works with countries to reverse the underinvestment in water, minimize the harm done by poor-quality water and maximize efficient water use. Alternatives to direct funding include ODA transfers, tariffs (to reduce leakages), user charges (which have socioeconomic implications), and taxes and subsidies. Countries can use ODA funds as seed money, generating savings that can be reinvested. First, however, countries need to build the capacity to secure the funds that are available: they need to know how to assemble knowledge and data, present a development plan and negotiate for funds.

Two other participants contributed to the discussion; one noted that the EC has funded many environmental health projects over recent decades, and identified WHO and the environment and health process as the main drivers. The public is increasingly aware of and concerned about environmental health issues. The challenge is to maintain funding, carry out impact assessment and convey the results to policy-makers, and identify and fund the best proposals. While climate change is a strong motive and attracts increasing amounts, it must not be allowed to divert funds from existing environmental health projects.

A participant raised the example of Serbia, where efforts to repair environmental damage include the investment of World Bank funds in an energy-efficient hospital in Belgrade. The funds are expected to be recouped in coming years, a prime example of how hospitals can lead the new wave of greening economies.

In addition, accession countries in south-eastern Europe are eligible for funding from the EU. They need to know where to apply, to build their capacity to negotiate for funds and to learn to spend them in a greener economy. This would not only bring in a return on the investment but also help to protect the environment. The public's increasing knowledge of the benefits of such projects could translate into support for policy-makers that pursue them.

World Health Youth Communication Network on Environment and Health: media awards

One of the key stakeholders in the environment and health debate is the mass media. The World Health Youth (WHY) Communication Network on Environment and Health comprises journalists aged 18–30 years from across the European Region. Designed to highlight the mass media's contribution to the environment and health debate, the WHY awards attracted over 40 entries, which addressed an environmental health issue of key importance to a country, evaluated its delivery in that country, illustrated some change over time and put a human face on the story. The applicants had to demonstrate writing ability in any multimedia form, show an interest in environmental health and have the support of their editors.

One of the founding members of the Network described the importance of identifying vested interests behind stories, separating news from advertising and learning to distinguish between evidence and opinion. The award winners confirmed their commitment to participating in the environment and health process as partners, stakeholders and resources. Five stories received the following prizes:

1. dental amalgam and the effects of mercury on the Danish environment;

2. the effects of climate change, such as drought, forest fires, heat waves and desertification in Spain;

3. low levels of water resources in Ukraine;

4. open-cast ore mines in Armenia;

5. the effects of climate change on rain, crops and food supply in Uzbekistan.

5. Dealing with climate change in Europe: challenges and synergies

Evidence confirms the increase in fossil fuel emissions and in hemispheric temperature. Modelling of climate change scenarios is developing, and paints a picture that must inform the response of health and environment experts and policies. The scenarios predict a significant increase in temperature with a significant decrease in precipitation, and countries need to adapt and mitigate where possible, even without knowing the effects of adaptation and being aware of the limitations of mitigation.

Health benefits of reducing greenhouse-gas emissions

Household energy use, urban land transport, the food and agriculture sector, and electricity generation result in large emissions of greenhouse gases. As mentioned, reducing these emissions would benefit health, as well as the environment. In housing and transport, preventing energy loss, improving the efficiency of fossil-fuel stoves and increasing physical activity can all prevent premature deaths and ill health, as well as reduce emissions. Deep cuts in emissions are needed; for example, the United Kingdom is estimated to need to make a reduction of 80% by 2020 for real change to result.

In the food and agriculture sector, 80% of emissions come from livestock production. The question is whether reducing the consumption of animal products is feasible, even though reducing the animal source of saturated fat by 30% could lower heart disease deaths by about 15% in the United Kingdom.

The production of low-carbon electricity needs to be reduced to 50% of its 2000 value by 2030 to make a real change. This reduction would also be associated with a decrease in acute and chronic effects of air pollution on health, particularly from PM, but the actual impact of this decrease is unclear.

In conclusion, lower-carbon strategies can save lives. These strategies need health impact assessments. The co-benefits for health can partly offset the costs of climate change mitigation, and this should be highlighted to ministries of finance. Health systems can lead this initiative by, for example, adopting low-carbon policies themselves.

Global political developments and health issues

In the context of the United Nations Framework Convention on Climate Change (UNFCCC), health is included in the area of adaptation, which is defined as "Adjustment in natural or human systems in response to actual or expected climatic stimuli or their effects, which moderates harm or exploits beneficial opportunities" *(33)*. This adjustment of human systems includes any response to health threats caused by climate change.

The Subsidiary Body for Scientific and Technological Advice for the UNFCCC, through its workshops in Nairobi, Kenya, has identified actions needed in the health sector:

"Climate change making an intervention of its own"

- promoting research and surveys of climate change impacts on health;

- mapping hazard and/or health vulnerability "hot spots";

- conducting community-based risk profiling, examining where risks to health from climate change are concentrated and using participatory research methods;

- developing geographical information system tools for risk reduction and response in cases of vector-borne disease;

- establishing surveillance and early warning systems for climate-related health risks;

- preparing guidelines for distribution through medical institutions and other actors to identify best practices and promoting training on risk assessment tools in the health sector and risk management techniques, including rapid assessment tools;

- developing a health strategy across agencies with a common portfolio of methods and tools;

- including climate risk information in existing health outreach activities;

- instituting programmes to identify long-term health needs, particularly in the context of disaster risk reduction; and

- developing indicators for tracking health risks and the effectiveness of adaptation actions.

The United Nations Climate Change Conference in December 2009 made some progress. In particular, it raised climate change policy to the highest political level. It significantly advanced the negotiations on long-term cooperative action, including on defining the functioning of the necessary infrastructures. The Copenhagen Accord *(34)* was an important political announcement, a clear message of political intent to constrain carbon emissions and respond to climate change in both the short and long terms.

The next steps for governments were to agree on a work programme for 2010, have informal consultations, consult and agree on approaches for future negotiations, and ensure the immediate operation of any arrangements. It is now possible to have some practical expectations of the Conference to be held in November–December 2010 in Mexico.

European Regional Framework for Action on climate change and health

The European Regional Framework for Action on climate change and health *(35)* was developed by a task force co-chaired by specialists from the United Kingdom and Serbia. The open-ended task force consisted of representatives of Member States, EC, EEA, ECDC, the Health and Environment Alliance and the Regional Environmental Center for Central and Eastern Europe. The Framework aims to protect health, promote health equity and security, and provide healthy environments in a changing climate in the WHO European Region. It has five strategic objectives:

- to ensure that health issues are integrated at all levels into all current and future measures, policies and strategies for climate change mitigation and adaptation;

- to strengthen health, social and environmental systems and services to improve their capacity to prevent, prepare for and cope with climate change;

- to raise awareness to encourage healthy mitigation and adaptation policies in all sectors;

- to increase the health and environment sectors' contribution to reducing greenhouse-gas emissions; and

- to share best practices, research, data, information, technology and tools at all levels on climate change, environment and health.

Countries' priorities for the next 20 years

A panel of country representatives outlined their priorities for the coming 20 years and how they intend to tackle them. Foremost among these were adaptation to extreme weather events, such as increasingly frequent and prolonged heat-waves; a number of countries are finalizing projects to ensure the intersectoral integration of health concerns in emergency preparedness plans. Other priorities are better mapping of the health effects of climate change and increased surveillance of vector-borne diseases. As to mitigation, countries are adopting a new approach characterized as the green economy, which includes greater energy efficiency, reduced pollution, more use of renewable energy sources and sustainable exploitation of natural resources. In terms of policy, countries focus on developing strategies and action plans on climate change and its impact on public health, although some are also looking at the effects on climate change caused by the health sector, such as emissions from hospitals.

In conclusion, health and economic development can indeed go hand in hand; synergies can be achieved in, for instance, the combined certification of energy efficiency and improved air quality. Reliance on sound science is a common feature of work in both health and the environment, and there is clearly a widespread willingness to take forward the Regional Framework for Action *(35)*.

Challenges and responses in the global health agenda

In a keynote address, Anarfi Asamoa-Baah, WHO Deputy Director-General, paid tribute to the vision and foresight of those who had conceived the European environment and health process. Largely thanks to European Member States and institutions, which championed the cause of the environment and health at a time when it was not fashionable, the challenges and responses in this area are now at the centre of the global health agenda. Nevertheless, continuing efforts must be made to strengthen the evidence base and deepen strategic alliances, notably with civil-society organizations, business, young people, and communication specialists and the media.

Although some familiar issues still need resolution – clean water supply and sanitation, for instance, are still a problem for the poorer segments of society – two main trends will dominate the future. One is the ageing of the population, 25% of whom will suffer from disability; the other is globalization, which entails easier travel not only for human beings but also for diseases and unhealthy lifestyles. Primary prevention and action on the social determinants of health, an approach that the WHO European Region is pioneering, are the best ways to meet the challenges resulting from these trends.

In a second keynote address, John Dalli, European Commissioner for Health and Consumer Policy, underlined the EC's strong support for WHO's European environment and health process. While the economic crisis, growing public debts and rises in unemployment dominated the headlines, health and the environment are key factors that underpin economic performance, recovery and success, in addition to people's well-being. Indeed, economic recovery cannot be sustained without a healthy population, and high environmental and health standards.

Environmental factors can significantly affect citizens' health and diseases' development and progression. Such factors can particularly affect vulnerable groups in society, such as children, pregnant women and socially disadvantaged people. For instance, one in every five children suffers from a chronic respiratory condition or allergy, so the need to address air quality is pressing, especially in indoor environments. A quarter of all European schoolchildren are overweight or obese; to reverse this trend, physical activity needs to be promoted, for example, by creating environments that encourage people to walk or cycle. Injuries, climate change, and water supply and sanitation all have massive implications for people's health. The EC and EU Member States have made reducing the social impact of the financial crisis – and thereby its health impact – a key priority.

At the Fourth Ministerial Conference in Budapest in June 2004, the EC presented its then newly adopted Environment and Health Action Plan *(36)*. The Action Plan was designed and implemented in close collaboration with WHO and in line with the pan-European process. Today, much of it has been implemented, with some funding from the EU Public Health Programme and the framework programmes for research. The Action Plan's main achievement has been to integrate the key policy areas of environment, health and research at the European level. The time has come to build on the progress made, focusing on two aspects:

- the integration of health and environment policies and the incorporation of health concerns in all policies; and

- work on solutions to ensure that the environment does not damage people's health.

This needs cooperation at the international, European and national levels, with the involvement of key NGOs and the business community. Naturally, international cooperation must be translated to the national level and focus on helping Member States take effective action. The EC is fully committed to working with WHO and its partners towards this end.

6. Future of the European environment and health process

The Conference resulted in two outcome documents: the Parma Declaration on Environment and Health and *The European environment and health process (2010–2016): institutional framework* (annexes 1 and 2).

Parma Declaration

Negotiations on the document began in 2007. To ensure active involvement, membership in the Declaration Drafting Group was open to all Member States; the members were Andorra, Armenia, Austria, Belgium, Croatia, Finland, France, Germany, Italy, the Netherlands, Norway, Serbia, Sweden, Turkey and the United Kingdom, as well as youth representatives, the International Trade Union Confederation, the Eco-Forum, the Health and Environment Alliance, the World Business Council for Sustainable Development, the EC, the Regional Environmental Center for Central and Eastern Europe, UNECE and WHO.

The document contains both a political declaration and a technical commitment to act. The political declaration consists of a plan of implementation through intensified effort; new challenges, such as climate change, new risks facing children and socioeconomic inequalities; and the need for effective mechanisms, such as better public services at the national level, work with different partners and sectors, and the funding to do it. There are needs to advocate investment in environmentally friendly and health-promoting technology, to implement the actions listed in the commitment to act and to strengthen collaboration through an institutional framework: a ministerial board at the political level and a task force at the implementation level, which will report to both WHO and UNECE. The next environment and health conference, planned for 2016, will follow up on progress.

The commitment to act section includes commitments in four areas: children's health and the RPGs; climate change; children, young people and other stakeholders; and the development of tools. The first area contains targets for achievement, notably children's access to safe water by 2020 and to healthy and safe environments to play in by 2020, clean indoor air free of tobacco smoke by 2015, a reduction in chemical risk by 2015 and the development of national plans to prevent asbestos-related diseases by 2015.

Action in the second area will protect health, well-being, natural resources and ecosystems and increase the health sector's contribution to reducing greenhouse-gas emissions. The implementation of the Regional Framework for Action *(35)* is recommended.

The third commitment is to involve children, young people and other stakeholders, not just Member States, in the process through youth participation, increased cooperation at the local and subnational levels, and building professional capacities. The fourth requires the further development of tools, such as ENHIS *(13)*, tools and guidelines on the economic impact of environmental health risks, and interdisciplinary tools for research on environment and health.

The three signatories of the Parma Declaration

Institutional framework

To ensure the future evolution of the European environment and health process, flexible, logical and sustainable structures for cross-sectoral, national and international collaboration are needed to strengthen the development and implementation of evidence-based policies. Annex 2 sets out the proposed institutional framework for the process.

Leading national officials from the health and environment sectors will meet annually at the regional level in the European Environment and Health Task Force, the leading regional body to implement and monitor the technical progress of the process. The Task Force will consist of representatives of Member States, as well as other key stakeholders and partners, such as the EC, the EU, United Nations agencies and NGOs. It will ensure communication and collaboration among stakeholders, at the national and international levels and between government and nongovernmental sectors, and review scientific evidence to advise on new challenges, policies and solutions.

The European Environment and Health Ministerial Board will sustain political commitment between ministerial conferences, and be accountable to the Member States through WHO and UNECE's existing governance mechanisms. The Board will consist of four ministers of health and four of the environment, and EC and United Nations representatives, for geographical and sectoral representation.

The Member States in the Region welcomed the Declaration, including the commitment to act, and the institutional framework for the environment and health process, adopting them by acclamation. The Conference participants agreed that environment and health issues are challenging, but the pressure of the financial crisis, climate change and other emerging threats make it imperative to act now. They supported the way the environment and health process has progressed, most clearly in the form of the two documents, while noting that some elements, such as the terms of reference of the Task Force, need further refinement. Most emphasized that political commitment was more important than the further clarification of details. They welcomed the flexibility, sustainability and transparency of the proposed framework and the inclusion of concrete targets in the commitment to act. They reiterated the importance of climate change, children's environmental health, the participation of young people, and socioeconomic and gender inequalities, while urging greater consideration of issues such as the environmental causes of chronic diseases, nano technology, and endocrine-disrupting and other hazardous chemicals. Some proposed having ad hoc thematic groups as needs arise, pointing out the benefits of sharing country examples, and some would prefer to have called for the substitution of asbestos.

A young people's delegation, working in parallel with and contributing regularly to the environment and health process, prepared a youth declaration, which they signed on behalf of the 73 youth delegates at the Conference and presented to the WHO Regional Director for Europe (Annex 3).

John Dalli, European Commissioner for Health and Consumer Policy, affirmed the importance that the EC attaches to the environment and health process and to working with all stakeholders to reduce the disease burden. He welcomed the Parma Declaration and its concrete steps for implementing the process. He acknowledged that, despite the progress made, addressing the goals remains a challenge but one that the EC is committed to meeting with its many partners (Annex 4).

The Minister of Environment, Land and Sea of Italy, the Minister of Health of Italy and the WHO Regional Director for Europe signed the Parma Declaration on behalf of all 53 Member States in the European Region and WHO.

Closing of the Conference

Stefania Prestigiacomo, Minister of Environment, Land and Sea of Italy, thanked all those who had contributed to the success of the Conference. She emphasized the value of communicating to the public the scientific basis for the links between environment and health. She underlined the great importance of the Parma Declaration, including its commitments to not only preventing environmental risks and diseases but also tackling emerging issues such as climate change and droughts, particularly their effect on children. She indicated that the broad consensus around the Declaration was a great achievement, especially coming soon after the 2009 United Nations Climate Change Conference, and undertook to take the process forward in Italy with the Minister of Health.

Ferrucio Fazio, Minister of Health of Italy, lauded the strong collaboration among all those involved in the Conference, indicating that it reflected the importance that the European Region attributes to environmental health. He referred to the coordination between the health and environment ministries in Italy, including in such areas as primary health care and disease prevention for children and elderly people, as an example of the work already taking place in the spirit of the Parma Declaration.

Closing the Conference, Zsuzsanna Jakab, WHO Regional Director for Europe, stated that the Conference had opened an exciting new chapter in the way European governments work on environment and health. In endorsing a new vision for the future of the European environment and health process, they had set new goals and commitments, and agreed on a new conceptual and operational framework. Through the Declaration, governments had committed to meet concrete targets in the next decade, to ensure:

- access to safe water and sanitation;

- opportunities for physical activity and a healthy diet;

- disease prevention through improved air quality; and

- healthy environments free of toxic chemicals.

She confirmed that progress would be monitored and evaluated very closely in the coming years.

The priority given to climate change and health opened up the possibility of more green jobs and more investment in new technologies, based on the Regional Framework for Action *(35)*. The health sector should now lead other sectors on reducing greenhouse-gas emissions, and work with the environment sector as advocates to other government sectors. Further, governments had pledged to reduce socioeconomic and gender inequalities in the human environment and health, and should find ways of targeting vulnerable groups and addressing the noncommunicable disease epidemic.

The involvement of ministers and a wide group of key stakeholders in the work of the European Environment and Health Ministerial Board and Task Force would strengthen the political and technical coordination required to succeed, as well as enhancing the status of public health in the Region. In this way, governments would be able to move closer to more just and equitable societies, by translating WHO's defining values of solidarity, equity and participation into action.

References

1. *Environment and health: the European Charter and commentary. First European Conference on Environment and Health, Frankfurt, 7–8 December 1989.* Copenhagen, WHO Regional Office for Europe, 1990 (WHO Regional Publications, European Series, No. 35).

2. *Environment and health: report on the second European conference, Helsinki, Finland, 20–22 June 1994.* Copenhagen, WHO Regional Office for Europe, 1995.

3. *Third Ministerial Conference on Environment and Health, London, 16–18 June 1999. Report.* Copenhagen, WHO Regional Office for Europe, 1999.

4. *Fourth Ministerial Conference on Environment and Health, Budapest, Hungary, 23–25 June 2004. Final conference report.* Copenhagen, WHO Regional Office for Europe, 2005.

5. *Children's Environment and Health Action Plan for Europe.* Copenhagen, WHO Regional Office for Europe, 2004 (http://www.euro.who.int/document/e83338.pdf, accessed 18 May 2010).

6. *Intergovernmental Midterm Review. Vienna 13–15 June 2007. Meeting report.* Copenhagen, WHO Regional Office for Europe, 2007 (http://www.euro.who.int/Document/EEHC/IMR_Vienna_mtgrep_en.pdf, accessed 18 May 2010).

7. *Health and environment in Europe. Progress assessment.* Copenhagen, WHO Regional Office for Europe, 2010 (http://www.euro.who.int/document/E93556.pdf, accessed 18 May 2010).

8. Transport, Health and Environment Pan-European Programme (THE PEP) [web site]. Geneva, United Nations Economic Commission for Europe, 2010 (http://www.unece.org/thepep/en/welcome.htm, accessed 27 May 2010).

9. Convention on the Protection and Use of Transboundary Watercourses and International Lakes. About the Protocol on Water and Health [web site]. Geneva, United Nations Economic Commission for Europe, 2010 (http://www.unece.org/env/water/text/text_protocol.htm, accessed 27 May 2010).

10. Protocol on Strategic Environmental Assessment (SEA) [web site]. Geneva, United Nations Economic Commission for Europe, 2010 (http://www.unece.org/env/eia/sea_protocol.htm, accessed 27 May 2010).

11. Convention on Long-range Transboundary Air Pollution [web site]. Geneva, United Nations Economic Commission for Europe, 2010 (http://www.unece.org/env/lrtap/, accessed 27 May 2010).

12. Commission on Social Determinants of Health. 2005–2008 [web site]. Geneva, World Health Organization, 2010 (http://www.who.int/social_determinants/thecommission/en/, accessed 27 May 2010).

13. European Environment and Health Information System (ENHIS) [web site]. Copenhagen, WHO Regional Office for Europe, 2010 (http://www.euro.who.int/en/what-we-do/data-and-evidence/environment-and-health-information-system-enhis, accessed 27 May 2010).

14. The United Nations Climate Change Conference in Copenhagen, 7–19 December 2009 [web site]. Bonn, United Nations Framework Convention on Climate Change, 2009 (http://unfccc.int/meetings/cop_15/items/5257.php, accessed 27 May 2010).

15. We can end poverty 2015. Millennium Development Goals [web site]. New York, United Nations, 2010 (www.un.org/millenniumgoals, accessed 27 May 2010).

16. *Declaration of the Third Ministerial Conference on Environment and Health*. Copenhagen, WHO Regional Office for Europe, 1999 (http://www.euro.who.int/en/who-we-are/policy-documents/declaration-of-the-third-ministerial-conference-on-environment-and-health, accessed 28 July 2010).

17. Convention on the Rights of the Child [web site]. New York, Unicef, 2008 (http://www.unicef.org/crc/, accessed 28 May 2010)

18. Strategic Approach to International Chemicals Management (SAICM) [web site]. Geneva, United Nations Environment Programme, 2010 (www.saicm.org, accessed 28 May 2010).

19. Commission on Social Determinants of Health – final report [web site]. Geneva, World Health Organization, 2008 (http://www.who.int/social_determinants/thecommission/finalreport/en/index.html, accessed 28 May 2010).

20. Global Health Equity Group. Strategic review of health inequalities in England post-2010 (Marmot review) [web site]. London, University College London, 2010 (http://www.ucl.ac.uk/gheg/marmotreview, accessed 28 May 2010).

21. Basic documents. Forty-seventh edition [web site]. Geneva, World Health Organization, 2009 (http://apps.who.int/gb/bd/, accessed 28 May 2010).

22. Healthier environments for children [web site]. Brussels, Health and Environment Alliance, 2010 (cehape.env-health.org, accessed 28 May 2010).

23. *Declaration of the Fourth Ministerial Conference on Environment and Health*. Copenhagen, WHO Regional Office for Europe, 2004 (http://www.euro.who.int/__data/assets/pdf_file/0008/88577/E83335.pdf, accessed 28 May 2010).

24. Morris GP et al. Getting strategic about the environment and health. *Public Health*, 2006, 120:889–907.

25. Regions for Health Network (RHN) [web site]. Copenhagen, WHO Regional Office for Europe, 2010 (http://www.euro.who.int/en/who-we-are/networks/regions-for-health-network-rhn, accessed 28 May 2010).

26. Global Strategy on Diet, Physical Activity and Health. Strategy documents [web site]. Geneva, World Health Organization, 2004 (http://www.who.int/dietphysicalactivity/strategy/eb11344/en/index.html, accessed 28 May 2010).

27. *European Charter on Counteracting Obesity*. Copenhagen, WHO Regional Office for Europe, 2006 (http://www.euro.who.int/__data/assets/pdf_file/0009/87462/E89567.pdf, accessed 28 May 2010).

28. *White paper on a strategy for Europe on nutrition, overweight and obesity-related health issues*. Brussels, Commission of the European Communities, 2007 (ec.europa.eu/health/ph_determinants/life…/nutrition_wp_en.pdf, accessed 28 May 2010).

29. The Eye on Earth [web site]. Copenhagen, European Environment Agency, 2010 (http://eyeonearth.cloudapp.net/, accessed 28 May 2010).

30. EU Platform on Diet, Physical Activity and Health [web site]. Brussels, Commission of the European Communities, 2010 (http://ec.europa.eu/health/nutrition_physical_activity/platform/index_en.htm, accessed 28 May 2010).

31. *Steps to health: a European framework to promote physical activity for health.* Copenhagen, WHO Regional Office for Europe, 2007 (http://www.euro.who.int/__data/assets/pdf_file/0020/101684/E90191.pdf, accessed 28 May 2010).

32. *Night noise guidelines for Europe.* Copenhagen, WHO Regional Office for Europe, 2009 (http://www.euro.who.int/__data/assets/pdf_file/0017/43316/E92845.pdf, accessed 28 May 2010).

33. Intergovernmental Panel on Climate Change. *Climate Change 2001: synthesis report. A contribution of working groups i, ii, iii to the third assessment report of the Intergovernmental Panel on Climate Change.* Cambridge and New York, Cambridge University Press, 2001 (http://www.ipcc.ch/ipccreports/tar/vol4/index.htm, accessed 28 May 2010).

34. Copenhagen Accord [web site]. Bonn, United Nations Framework Convention on Climate Change, 2010 (http://unfccc.int/home/items/5262.php, accessed 28 May 2010).

35. *Protecting health in an environment challenged by climate change. European Regional Framework for Action.* Copenhagen, WHO Regional Office for Europe, 2010 (http://www.euro.who.int/__data/assets/pdf_file/0005/95882/Parma_EH_Conf_edoc06rev1.pdf, accessed 28 May 2010).

36. *The European Environment & Health Action Plan 2004–2010.* Brussels, Commission of the European Communities, 2004 (http://ec.europa.eu/environment/health/pdf/com2004416.pdf, accessed 28 May 2010).

Annex 1. Parma Declaration on Environment and Health and commitment to act

Parma Declaration on Environment and Health

1. We the Ministers and Representatives of Member States in the European Region of the World Health Organization (WHO) responsible for health and the environment, together with the WHO Regional Director for Europe, in the presence of the European Commissioners for Health and Consumer Policy and for the Environment, the Executive Secretary of the United Nations Economic Commission for Europe (UNECE) and the Regional Director for Europe of the United Nations Environment Programme (UNEP) have gathered in Parma, Italy from 10 to 12 March 2010 to face the key environment and health challenges of our time.

2. Building on the foundations laid in the European Environment and Health Process to date, we will intensify our efforts to implement the commitments made through previous WHO ministerial conferences, especially those set out in the Children's Environment and Health Action Plan for Europe (CEHAPE).

3. We are committed to act on the key environment and health challenges of our time. These include:

    (a) the health and environmental impacts of climate change and related policies;

    (b) the health risks to children and other vulnerable groups posed by poor environmental, working and living conditions (especially the lack of water and sanitation);

    (c) socioeconomic and gender inequalities in the human environment and health, amplified by the financial crisis;

    (d) the burden of noncommunicable diseases, in particular to the extent that it can be reduced through adequate policies in areas such as urban development, transport, food safety and nutrition, and living and working environments;

    (e) concerns raised by persistent, endocrine-disrupting and bio-accumulating harmful chemicals and (nano)particles, and by novel and emerging issues; and

    (f) insufficient resources in parts of the WHO European Region.

4. We will address these challenges by setting up or strengthening existing mechanisms or structures that can ensure effective implementation, promote local actions and ensure active participation in the European Environment and Health Process. Recognizing that economic arguments are increasingly critical to develop sound policies, we will pay special attention to fostering strategic partnerships and networks, so that environment and health issues are better integrated across the policies of all sectors. We call on these sectors and relevant organizations to work with us more closely to ensure healthy environments.

5. We will intensify efforts to develop, improve and implement health and environmental legislation and to continue health system reforms as necessary, particularly in the newly independent states and countries

of south-eastern Europe, aimed at streamlining, upgrading and strengthening the performance of public health and environmental services.

6. We will ensure that youth participation is facilitated across all Member States at both national and international levels by providing them with assistance, resources and the training required for meaningful and sustainable involvement in all aspects of the process.

7. We will advocate for investing in sustainable and environmentally friendly and health-promoting technologies, emphasizing the opportunities created by these activities, such as energy-efficient health services and green jobs.

8. We encourage international stakeholders, including international financial institutions, and the European Commission to offer further scientific, political, technical and financial assistance to help establish effective mechanisms and strengthen capacities to reduce exposures to environmental hazards and the resulting health impacts in the Region.

9. We call upon the WHO Regional Office for Europe, the European Commission, UNECE, UNEP and all other partners to strengthen their collaboration to ensure progress in environment and health implementation in the WHO European Region.

10. We endorse and will implement the "Commitment to act" and the goals and targets included therein. That document is an integral part of this Declaration.

11. We endorse the institutional framework described in the "The European Environment and Health Process (2010–2016): Institutional framework". We commend a stronger political role for the European Environment and Health Ministerial Board and we will follow up on implementation through the Environment and Health Task Force and the Ministerial Board will report annually to the WHO Regional Committee for Europe and the UNECE Committee on Environmental Policy.

12. We agree to meet again at the Sixth European Ministerial Conference on Environment and Health in 2016.

13. We the Minister of Health and the Minister of the Environment, Land and Sea of Italy, on behalf of all the ministers of health and environment in the European Region of WHO, together with the WHO Regional Director for Europe and in the presence of the European Commissioners for Health and the Environment, the Executive Secretary of UNECE and other partners, hereby fully adopt the commitments made in this Declaration.

Minister of Health, Italy
Co-president

Minister of Environment, Italy
Co-president

WHO Regional Director
for Europe

Commitment to act

Building on the foundations laid in the European Environment and Health Process to date, including in particular the Fourth Ministerial Conference on Environment and Health and the Intergovernmental Mid-term Review held in Vienna in June 2007, we will increase our efforts to address the key environment and health challenges of our time, including climate change, emerging issues and the effects of the economic crisis, and we reaffirm our commitment to work together across sectors.

We recognize established political processes that ensure healthy environments for children, including all related United Nations processes, other WHO ministerial conferences as well as European Union legislation and the 2009 deliberations of the Group of Eight industrialized nations (G8), as tools for further implementation.[1]

We take particular note of the Declaration of the Sixth Ministerial Conference "Environment for Europe", of WHO's Tallinn Charter on Health Systems, Health and Wealth[2] and of the European Union Declaration on Health in All Policies.

A. Protecting children's health

1. We reconfirm our commitment to prioritized actions under the regional priority goals (RPGs) in the Children's Environment and Health Action Plan for Europe (CEHAPE) as indicated below. We will strive to attain the targets in the RPGs as set out below.

Regional Priority Goal 1. Ensuring public health by improving access to safe water and sanitation

   i. We will take advantage of the approach and provisions of the Protocol on Water and Health[3] as a rationale and progressive tool to develop integrated policies on water resource management and health, addressing the challenges to safe water services posed by climate change, with clear targets and objectives, working in partnership with all concerned sectors.

   ii. We will strive to provide each child with access to safe water and sanitation in homes, child care centres, kindergartens, schools, health care institutions and public recreational water settings by 2020, and to revitalize hygiene practices.

Regional Priority Goal 2. Addressing obesity and injuries through safe environments, physical activity and healthy diet

   i. We will implement the relevant parts of the commitments set out in the Amsterdam Declaration of the Third High-Level Meeting of the Transport Health and Environment Pan-European Programme (THE PEP).

   ii. We will integrate the needs of children into the planning and design of settlements, housing, health care institutions, mobility plans and transport infrastructure. To this end we will use health, environment and strategic impact assessments and we will develop and adapt the relevant regulations, policies and guidelines, and implement the necessary measures.

---

[1] Turkey declares that it does not consider itself bound by the commitments and undertakings in the paragraphs related to international treaties, conventions or protocols to which it is not a contracting party, namely the Protocol on Water and Health to the 1992 Convention on the Protection and Use of Transboundary Watercourses and International Lakes and the Protocols to the 1979 Convention on Long-Range Transboundary Air Pollution except the 1984 Protocol on Long-Term Financing of the Cooperative Programme for Monitoring and Evaluation of the Long-range Transmission of Air Pollutants in Europe.

[2] Within the political and institutional framework of each country, a health system is the ensemble of all public and private organizations, institutions and resources mandated to improve, maintain or restore health. Health systems encompass both personal and population services, as well as activities to influence the policies and actions of other sectors to address the social, environmental and economic determinants of health.

[3] Protocol on Water and Health to the 1992 Convention on the Protection and Use of Transboundary Watercourses and International Lakes.

iii. We will work in partnership with local, regional and national authorities to advocate for actions to counteract the adverse effects of urban sprawl that cause socioeconomic, health and environmental consequences.

iv. We aim to provide each child by 2020 with access to healthy and safe environments and settings of daily life in which they can walk and cycle to kindergartens and schools, and to green spaces in which to play and undertake physical activity. In so doing, we intend to prevent injuries by implementing effective measures and promoting product safety.

v. We will implement the WHO European Action Plan for Food and Nutrition Policy (2007–2012), in particular by improving the nutritional quality of school meals, and support local food production and consumption, where it can reduce environmental and health impacts.

Regional Priority Goal 3. Preventing disease through improved outdoor and indoor air quality

i. We will take advantage of the approach and provisions of the protocols to the 1979 Convention on Long-Range Transboundary Air Pollution and we will support their revision, where necessary. We will continue and enhance our efforts to decrease the incidence of acute and chronic respiratory diseases through reduction of exposure to ultrafine particles and other particulate matter, especially from industry, transport and domestic combustion, as well as ground-level ozone, in line with WHO's air quality guidelines. We will strengthen monitoring, control and information programmes, including those related to fuels used in transport and households.

ii. We will develop appropriate cross-sectoral policies and regulations capable of making a strategic difference in order to reduce indoor pollution, and we will provide incentives and opportunities to ensure that citizens have access to sustainable, clean and healthy energy solutions in homes and public places.

iii. We aim to provide each child with a healthy indoor environment in child care facilities, kindergartens, schools and public recreational settings, implementing WHO's indoor air quality guidelines and, as guided by the Framework Convention on Tobacco Control, ensuring that these environments are tobacco smoke-free by 2015.

Regional Priority Goal 4. Preventing disease arising from chemical, biological and physical environments

i. We will take advantage of the approach and provisions of relevant international agreements.[4] We will contribute to the Strategic Approach to International Chemicals Management (SAICM) and to the development of the global legal instrument on mercury.

ii. We aim to protect each child from the risks posed by exposure to harmful substances and preparations, focusing on pregnant and breast-feeding women and places where children live, learn and play. We will identify those risks and eliminate them as far as possible, by 2015.

iii. We will act on the identified risks of exposure to carcinogens, mutagens and reproductive toxicants, including radon, ultraviolet radiation, asbestos and endocrine disruptors, and urge other stakeholders to do the same. In particular, unless we have already done so, we will develop by 2015 national programmes for elimination of asbestos-related diseases in collaboration with WHO and ILO.

---

[4] Such as the Basel Convention on the Control of Transboundary Movements of Hazardous Wastes and their Disposal, the Rotterdam Convention on the Prior Informed Consent Procedure for Certain Hazardous Chemicals and Pesticides in International Trade, and the Stockholm Convention on Persistent Organic Pollutants, as well as the protocols on heavy metals and on persistent organic pollutants to the 1979 Convention on Long-Range Transboundary Air Pollution.

iv. We call for more research into the potentially adverse effects of persistent, endocrine-disrupting and bio-accumulating chemicals and their combination, as well as for the identification of safer alternatives. We also call for an increase of research into the use of nanoparticles in products and nanomaterials, and electromagnetic fields, in order to evaluate possible harmful exposures. We will develop and use improved health risk and benefit assessment methods.

v. We call upon all stakeholders to work together to reduce children's exposure to noise, including that from personal electronic devices, recreation and traffic, especially in residential areas, at child care centres, kindergartens, schools and public recreational settings. We urge and offer our assistance to WHO to develop suitable guidelines on noise.

vi. We will pay particular attention to child labour and exploitation as one of the major settings of exposure to relevant risks, and especially to hazardous chemicals and physical stressors.

B. Protecting health and the environment from climate change

2. We are committed to protecting health and well-being, natural resources and ecosystems and to promoting health equity, health security and healthy environments in a changing climate. Taking into account the ongoing work under the United Nations Framework Convention on Climate Change and recognizing subregional, socioeconomic, gender and age variability, we will:

   i. integrate health issues in all climate change mitigation and adaptation measures, policies and strategies at all levels and in all sectors. We will assess, prevent and address any adverse health effects of such policies by, for example, strengthening health promotion in environmental policies;

   ii. strengthen health, social welfare and environmental systems and services to improve their response to the impacts of climate change in a timely manner, for example to extreme weather events and heat waves. In particular, we will protect the supply of water and the provision of sanitation and safe food through adequate preventive, preparedness and adaptive measures;

   iii. develop and strengthen early warning surveillance and preparedness systems for extreme weather events and disease outbreaks, for example vector-borne diseases, at the animal-human-ecosystem interface, where appropriate;

   iv. develop and implement educational and public awareness programmes on climate change and health, to encourage healthy, energy-efficient behaviours in all settings and provide information on opportunities for mitigation and adaptation interventions, with a particular focus on vulnerable groups and subregions;

   v. collaborate to increase the health sector's contribution to reducing greenhouse gas emissions and strengthen its leadership on energy- and resource-efficient management and stimulate other sectors, such as the food sector, to do the same;

   vi. encourage research and development, for example with tools for forecasting climate impacts on health, identifying health vulnerability and developing appropriate mitigation and adaptation measures.

3. We call on the WHO Regional Office for Europe, to discuss with the European Commission, the European Environment Agency, the United Nations Economic Commission for Europe, the United Nations Environment Programme and other partners, on setting up European information platforms for systematic sharing of best practices, research, data, information, technology and tools focused on health at all levels.

4. We welcome the Regional Framework for Action entitled *Protecting health in an environment challenged by climate change*. We recommend that the approaches described in it are used to support action in this area.

C. Involvement of children, young people and other stakeholders

5. We will ensure that youth participation in national as well as international processes is facilitated across all Member States by providing them with assistance, adequate resources and the training required, and by giving them opportunities for meaningful involvement.

6. We will increase our cooperation with local and subnational authorities, intergovernmental and nongovernmental organizations, the business community, trade unions, professional associations and the scientific community, drawing on their experience and knowledge in order to achieve the best possible results.

7. We call on the business community to address the challenges posed in this commitment, for instance through relevant corporate and sectoral programmes.

8. We will seek to improve knowledge of environment and health issues and build the capacity of all professionals, with particular emphasis on health professionals and professional caretakers of children.

D. Knowledge and tools for policy-making and implementation

9. We support the development of the European Environment and Health Information System (ENHIS). We call on the WHO Regional Office for Europe, and also on the European Commission and the European Environment Agency to continue to assist Member States with the development of internationally comparable indicators, and to assist in the interpretation and practical application of relevant research results.

10. We encourage all relevant international organizations to further develop common tools and guidelines to address the economic impacts of environmental risk factors to health, including the cost of inaction, thereby facilitating the development and enforcement of legal instruments.

11. We will contribute to develop a consistent and rational approach to human biomonitoring as a complementary tool to assist evidence-based public health and environmental measures, including awareness-raising for preventive actions.

12. We acknowledge the contributions, conclusions and recommendations of the International Public Health Symposium on Environment and Health Research held in Madrid in October 2008. We agree to secure support for interdisciplinary research in line with the policy objectives of this Declaration and to improve the development of identified tools,[5] including health impact assessment. We will use existing information for policy-making and apply the precautionary principle where appropriate, especially in respect of new and emerging issues.

13. We affirm the need for participation of the public and stakeholders in tackling environment and health issues. We will develop and implement initiatives on risk perception, assessment, management and communication.

---

[5] Such as the Protocol on Strategic Environmental Assessment to the Convention on Environmental Impact Assessment in a Transboundary Context.

Annex 2. The European environment and health process (2010–2016): institutional framework

Introduction

The European environment and health process (EEHP) will continue towards the Sixth Ministerial Conference on Environment and Health in 2016. To ensure appropriate coordination between national implementation and international policies, an institutional framework is being proposed to ensure the proper level of monitoring and implementation, as well as political drive.

National mechanisms and structures

Member States are urged to set up or strengthen existing national environment and health mechanisms most appropriate to their specific national circumstances, to ensure implementation of the Parma Ministerial Conference commitments.

The European Environment and Health Task Force

The European Environment and Health Task Force (EHTF) will be the leading international body for implementation and monitoring of the EEHP.

Composition

The EHTF will include leading officials from the national implementation mechanisms and structures of the 53 Member States in the WHO European Region, nominated at national level as focal points for the EEHP.

WHO, the European Commission, the United Nations Environment Programme (UNEP), the United Nations Economic Commission for Europe (UNECE), the United Nations Development Programme (UNDP), the Organisation for Economic Co-operation and Development (OECD), the European Environment Agency (EEA), the European Centre for Disease Prevention and Control (ECDC), the World Business Council for Sustainable Development, the International Trade Union Confederation, the Regional Environmental Center, the Health and Environment Alliance, the European Eco-Forum and the Environment and Health Youth Network will be full members of the EHTF.

Terms of reference

The EHTF will:

- provide a forum for exchange of technical experience and knowledge through discussion and exchange of good practice;

- regularly review scientific evidence with the support of WHO, UNECE, UNEP and other relevant institutions in order to encourage Member States to update, modify or strengthen existing policies, as appropriate;

- facilitate collaboration among relevant sectors, partners and stakeholders, including intergovernmental organizations, nongovernmental organizations, trade unions, the business community, young people, technical agencies and international financial institutions;

- promote specific initiatives on emerging issues;

- establish ad hoc working groups, task forces and other bodies, as necessary, on a temporary basis to address specific needs and issues;

- collaborate closely with the European Environment and Health Ministerial Board.

Method of work

The EHTF will meet annually in the period leading up to the Sixth Ministerial Conference on Environment and Health in 2016. One of the EHTF meetings will be a high-level mid-term meeting to review progress in implementation of the EEHP and its institutional framework, which will be convened no later than 2014. This high-level meeting will also be attended by deputy ministers or state secretaries, chief medical officers and senior environmental administrators.

The EHTF will be led by a chairperson and a co-chairperson, one from the health sector and one from the environment sector, elected at the first meeting. The chairpersons will hold their seat for one year only, in order to ensure rotation among as many countries as possible in the six-year span of the Task Force. To ensure continuity of the process, at the end of his/her one-year term of office, the co-chairperson will become the chairperson, and a new co-chairperson will be elected. The outgoing chairperson will be called upon to provide advice to the chairperson and the new co-chairperson, as required, forming a sort of "EHTF troika".

The European Environment and Health Ministerial Board

The European Environment and Health Ministerial Board (EHMB) will be the political face and driving force of international policies in the field of environment and health for implementation of the commitments made within the EEHP.

Composition

The EHMB will consist of eight ministers or their high-level representatives appointed by the WHO Regional Committee for Europe for the health sector and UNECE Committee on Environmental Policy for the environment sector, in a way that ensures geographic representation of all parts of the WHO European Region and equal representation of the health and environment sectors. Ministers will serve a two-year term of office.

Other members of the Ministerial Board will include the WHO Regional Director for Europe, the Executive Secretary of UNECE, the Director of the UNEP Regional Office for Europe and the European Commission.

During their tenure, the chairperson and co-chairperson of the EHTF will be members of the EHMB, to ensure close links between the two bodies.

Terms of reference

The EHMB will:

- put the EEHP into a broad public health and environment agenda;

- review and propose policy directions and strategic priorities;

- advocate further development of environment and health policies;

- identify financial opportunities that would enable implementation where resources are lacking;
- reach out to other sectors and stakeholders;
- collaborate closely with the EHTF.

Method of work

The EHMB will select two co-chairs from among its members.

The EHMB will be accountable to the WHO Regional Committee for Europe and the UNECE Committee on Environmental Policy (CEP). The EHMB will develop its agenda and role to ensure the political relevance and effective leadership of the whole EEHP. Its annual meetings will be arranged back-to-back with sessions of the Regional Committee and meetings of the CEP in alternating years, to facilitate attendance and to ensure the link to the two bodies that are its source of legitimacy.

Secretariat

The whole institutional framework will be serviced by the WHO Regional Office for Europe, which will cooperate closely with UNECE and the UNEP Regional Office for Europe.

Conclusion and next steps

The WHO Regional Committee for Europe, at its sixtieth session to be held in Moscow in September 2010, will be asked to take the necessary action to endorse the outcomes of the Fifth Ministerial Conference on Environment and Health. The UNECE Committee on Environmental Policy, at its meeting in October 2010, will be asked to do the same. The WHO Regional Committee for Europe will appoint members of the EHMB from the health sector, while members from the environment sector will be appointed by UNECE's CEP.

Before the end of 2010, Member States will be asked to nominate focal points for the EEHP, to provide an operational network for continuous collaboration between Member States and to attend the first meeting of the EHTF, to be held no later than June 2011. This will ensure a quick start to implementation of the political outcomes of the Fifth Ministerial Conference and timely reporting back to the meeting of the EHMB to be held back-to-back with the sixty-first session of the WHO Regional Committee for Europe in 2011.

The EHTF and the EHMB will elaborate their terms of reference and develop their rules of procedure in line with the present document.

Annex 3. Parma Youth Declaration 2010

Preamble

We, the young people participating at the Fifth Ministerial Conference on Environment and Health, Parma, Italy, 10–12 March 2010, have as our main concern the future of our health and the lives of future generations. Our declaration underlines the measure of our concern about the impact the environment has upon our health. Outcomes of the decisions taken at this Conference will be our inheritance, so we commit ourselves to following its outcomes and calling our policy-makers to account for their actions. Through our network we will build strong structures throughout the Region and dedicate ourselves to work with our governments and partners in the implementation of good policy and to challenge those policies that are weak and where we see slow progress in implementation.

Youth participation

1.0 Our participation in the planning and implementation of environment and health policy and strategy is not negotiable. Neither is our health or access to a clean and safe environment. We look forward to the continuation and strengthening of our participation in the new structure of the European Environment and Health Committee.

1.1 All policy sectors must work together in creating policy that is robust and sustainable, remembering the health and well-being of future generations.

1.2 Through our transparent and democratic network we will be part of local, regional and national meetings and strengthen our role in international meetings.

Education

2.0 The role of education in understanding the need to protect and sustain our environment is essential. Environmental education has to be an integral part of every child and young person's life, both through formal and non-formal curricula. Education does not end when we leave school.

2.1 Education about how to create and live in a healthy, safe and sustainable environment needs to be on the agendas of governmental, nongovernmental and international organizations.

Protecting children's health through the CEHAPE Regional Priority Goals

RPG1. Ensure safe water and sanitation

3.0   Access to clean water is the most basic human need. In some parts of the Region we still find homes and schools without clean running water. This is a gross inequality. We expect technologies to be used to improve access to water, conserve water use and enhance sanitation. We urge governments to work with us in finding solutions to the impact of climate change on our future water supply and to the use of science and technology to support our efforts.

RPG2. Ensure protection from injuries and adequate physical activity (obesity, mobility, injuries and urban green spaces)

3.1   Obesity is a risk to our health, active life and life expectancy. Tackling obesity requires a multidisciplinary approach including education, lifestyles and living conditions. We call upon our governments to work with us to establish programmes that help us provide safe and healthy nutrition, effective education and physical activity.

3.2   Injuries are the greatest killer of children in Europe. We believe this is an unacceptable factor in our environment it is therefore imperative that Member States create safer daily living conditions for children throughout Europe. The most effective way to achieve this is consultation with us and combined education, training and enforcement strategies.

RPG3. Ensure clean outdoor and indoor air

3.3   As children and young people, we are more sensitive to both indoor and out door air pollutants such as tobacco, industrial and transport emissions. Nine out of ten of people living in urban areas are subject to unsafe levels of out door air pollution. We expect more far-reaching international action and collaboration to fix this problem.

3.4   Banning smoking in public places has to be a pan-European policy. Nonsmokers have the right to live in a smoke-free environment, especially children and young people.

RPG4. Ensure a chemical and biological risk free environment

3.5   We do not know the risks that many new technologies, including genetically modified organisms and nano technologies, pose to our health and environment. It is imperative that independent expert research is undertaken and publically published before these technologies are put to use.

3.6   We congratulate countries that have removed asbestos from all products and materials and expect those who have not, to have done so by 2015.

3.7   Canada has banned the use of bisphenol A in baby bottles due to health concerns. We strongly feel that Member States should follow their lead.

Climate change

4.0   We believe that dangerous aspects of climate change pose one of the most significant long-term health threats to the people of Europe; therefore we propose limiting greenhouse-gas emissions to prevent a 1.5 °C average temperature increase on post-development temperatures. Funds need to be provided to ensure that less developed states meet this goal. Moreover, we see the need for the adaptation of health systems responding to emergent problems posed by climate change.

4.1   We note with mounting frustration the inability of states to negotiate a real solution to climate change and expect our Member States to act responsibly. Furthermore, climate change has links to other important issues including green taxation and deforestation.

The future

5.0   We welcome the collaboration that has been set up between our network, WHO, the European Commission, governments and other bodies. We shall continue to provide the youth perspective and ideas on environment and health issues in Europe.

5.1   We will build on the foundations as set out in section 6 of the 2010 Parma Declaration, which states that: "We will ensure that youth participation is facilitated across all Member States at both national and international levels by providing them with assistance, resources and the training required for meaningful and sustainable involvement in all aspects of the process". We look forward to the realization of this clause, ensuring our organization will participate at the highest and widest level possible in the environment and health process. Moreover, we will work closely with ministries of health and ministries of environment to develop collaborative and meaningful participatory partnerships at all levels, assisting them to fulfil this commitment to youth participation.

Signed, 12 March 2010

[signature]

WHO CEHAPE Youth Network Representative
For and on behalf of the WHO CEHAPE Youth Network

Note. All evidence used is from *Children's health and the environment in Europe: a baseline assessment.* Copenhagen, WHO Regional Office for Europe, 2007 (http://www.euro.who.int/__data/assets/pdf_file/0009/96750/E90767.pdf, accessed 28 May 2010).

Annex 4. Declaration of the European Commission[1]

The European Commission welcomes the renewed commitment to strengthen the links between environment and health set out in the final declaration of the Ministerial Conference held in Parma on 10–12 March 2010. The document sets out clearly how to implement the environment and health process across Europe, based *inter alia* on the Children's Environment and Health Action Plan for Europe (CEHAPE).

The Commission supports the focus given to key environment and health challenges, such as the impact of climate change on health and the environment, socioeconomic and gender inequalities and the burden of noncommunicable diseases linked to environmental conditions and disasters.

Despite the progress achieved so far, addressing major preventable determinants and diseases, such as obesity, respiratory and cardiovascular diseases, remains a major challenge. Renewed efforts are required to properly address the environmental causes of such determinants and diseases. The Parma ministerial declaration provides a strong impetus to the implementation of effective actions to further reduce the burdens that arise from such diseases.

The Commission is committed to working with governments, civil society and with international organizations, in particular the World Health Organization, to support as appropriate the achievement of the goals set out in the Parma Declaration.

In implementing its European Environment and Health Strategy[2] through the European Union (EU) Action Plan on Environment and Health,[3] the Commission will ensure that synergies between EU-level actions and those arising from the Parma Conference are fully exploited.

Signed on behalf of the European Commission

John Dalli

Parma, 12 March 2010

---

[1] Reproduced with permission of the European Commission (http://ec.europa.eu/health/healthy_environments/docs/parma_declaration_en.pdf).
[2] COM (2003) 338 final of 11.6.2003.
[3] COM (2004) 416 final of 9.6.2004.

Annex 5. Programme

Opening of the Conference

Pietro Vignali, Lord Mayor of Parma
Vincenzo Bernazolli, President of the Province of Parma
Stefania Prestigiacomo, Minister of Environment, Land and Sea of Italy
Ferruccio Fazio, Minister of Health of Italy
Zsuzsanna Jakab, WHO Regional Director for Europe
Ján Kubiš, Executive Secretary, United Nations Economic Commission for Europe
Margaret Chan, WHO Director-General (by video address)

Session 1. Environment and health in Europe – an assessment of progress

Chairs: Corrado Clini and Jon Hilmar Iversen, European Environment and Health Committee

Progress in environment and health in Europe, 1989–2010
Michal Krzyzanowski, WHO Regional Office for Europe

Environment and health at the global level: progress and challenges
Maria Neira, Director, Public Health and Environment, WHO headquarters

Ten years of the Water and Health Protocol – main achievements
Gheorghe Constantin, General Director, General Directorate for Water Management, Romania

The Transport, Health and Environment Pan-European Programme – making the difference
Julie Ng-A-Tham, Chair, THE PEP Steering Committee

Progress in protecting human health and the environment through environmental agreements and cooperation
Christophe Bouvier, Director, United Nations Environment Programme Regional Office for Europe

Discussion panel

Robert Thaler, Austria
Mihály Kökény, Hungary
Roberto Bertollini, WHO headquarters

Session 2. Environment and health challenges in a globalized world: socioeconomic and gender inequalities – why do they matter?

Chairs: Georgia and Malta

Health inequalities in Europe: role of environmental determinants
Sir Michael Marmot, Director, International Institute for Society and Health, London, United Kingdom

Equity in health, climate and the environment
Giovanni Berlinguer, Professor of Occupational Health, University of Rome, Italy, and Member of the WHO Commission on Social Determinants of Health

Discussion panel

Germany
Malta
Russian Federation
Slovenia

Session 3. Children's Environment and Health Action Plan for Europe (CEHAPE) awards

Chairs: Sascha Gabizon, Women in Europe for a Common Future (WECF), and Génon K. Jensen, Health and Environment Alliance (HEAL)

Awards ceremony

Discussion panel

Armenia
Austria
Azerbaijan
Belgium
Denmark
Netherlands
Norway
European Commission, Directorate-General for Health and Consumers

Implementing the Children's Environment and Health Action Plan for Europe (CEHAPE)

Chairs: Kyrgyzstan and Hungary

Austrian Children's Environment and Health Action Plan: Implementation makes a difference
Reinhard Mang, Secretary General, Federal Ministry of Agriculture, Forestry, Environment and Water Management, Austria

Implementing CEHAPE: regional overview of achievements and challenges
George Morris, Consultant, Ecological Public Health, NHS Health Scotland, United Kingdom

Discussion panel

Cyprus
Poland

Portugal
Republic of Moldova
Regional Environmental Center for Central and Eastern Europe
Youth representative

Session 4. Investing in environment and health: working with partners and stakeholders

Chairs: Norway and Ukraine

Working with partners and stakeholders
Gaudenz Silberschmidt, Deputy Director, International Relations Department, Ministry of Health, Switzerland

Partnerships in public health. Lessons from a region
Göran Henriksson, Västra Götaland Region of Sweden, Regions for Health Network

Engaging stakeholders: the European Food Safety Authority perspective
Catherine Geslain-Lanéelle, Executive Director, European Food Safety Authority

Discussion panel

Bulgaria
France
International Trade Union Confederation
Lord Mayor of Parma
World Business Council
Youth representative

Investing in environment and health: role of international financing mechanisms

Chairs: Montenegro and Portugal

Environmental finance for structural transformation
Kori Udovički, Regional Director, United Nations Development Programme Regional Bureau for Europe and the Commonwealth of Independent States

Discussion panel

Albania
Tajikistan
European Centre for Disease Prevention and Control
Organisation for Economic Co-operation and Development

World Health Youth (WHY) Communication Network on Environment and Health media awards

Master of ceremonies: Franklin Apfel

The importance of engaging media and especially young journalists from Budapest to Parma and beyond
Deborah Cohen, Senior Health Editor, *British Medical Journal*

Awards ceremony

Session 5. Dealing with climate change in Europe. Challenges and synergies

Chairs: Germany and Serbia

Climate change: challenges ahead for the European Region
Antonio Navarra, National Institute of Geophysics and Volcanology, Bologna, Italy

Public health benefits of strategies to reduce greenhouse gas emissions
Sir Andy Haines, Director, London School of Hygiene and Tropical Medicine, United Kingdom

Climate change: global political developments and health-related issues
Wanna Tanunchaiwatana, United Nations Framework Convention on Climate Change

Introducing the European framework for action on climate change and health
David Harper, Department of Health, England, United Kingdom

Discussion panel

Croatia
Estonia
Kyrgyzstan
Spain
The former Yugoslav Republic of Macedonia

Keynote address

Positioning environment and health challenges and responses within the global health agenda. The way forward
Anarfi Asamoa-Baah, Deputy Director-General, World Health Organization

Keynote address

John Dalli, European Commissioner for Health and Consumer Policy, European Commission

Session 6. The future of the European environment and health process

Global to local environment and health information
Jacqueline McGlade, Executive Director, European Environment Agency

Presentation of the ministerial declaration
Leen Meulenbergs, Chair, Declaration Drafting Group

Vision for the future of the European environment and health process
Zsuzsanna Jakab, WHO Regional Director for Europe

Signing of the Conference Declaration and closure of the Conference

Stefania Prestigiacomo, Minister of Environment, Land and Sea of Italy
Ferruccio Fazio, Minister of Health of Italy
Zsuzsanna Jakab, WHO Regional Director for Europe

Annex 6. Core publications

The complete documentation from the Conference is available in hard copy from the WHO Regional Office for Europe and in electronic form on its web site.[1]

Working documents

Parma Declaration on Environment and Health and commitment to act
Protecting health in an environment challenged by climate change: European Regional Framework for Action
The European environment and health process (2010–2016): institutional framework

Policy briefs

Social and gender inequalities in environment and health
Specific needs of the newly independent states and the countries of south-eastern Europe
Climate change and health in Europe: opportunities for action in partnership
The future of the European environment and health process

Background documents

Health and environment in Europe. Progress assessment
The journey to Parma: a tale of 20 years of environment and health action in Europe
Progress and challenges on water and health: the role of the Protocol on Water and Health

---

[1] Documentation [web site]. Copenhagen, WHO Regional Office for Europe, 2010 (http://www.euro.who.int/en/home/conferences/fifth-ministerial-conference-on-environment-and-health/documentation, accessed 1 June 2010).

Annex 7. Pre-Conference and side events

On 9 March 2010, the day before the Conference opened, a Water and Health Protocol Day was held to illustrate the five main pillars of the Protocol and demonstrate its role in and relevance to the environment and health process. A media workshop and a youth event also took place.

Side events during the Conference included symposia, standing coffees and poster sessions.

Pre-Conference events (9 March 2010)

Water and Health Protocol Day

The UNECE/WHO Regional Office for Europe Protocol on Water and Health: where health, environment and development policies meet

The five thematic sessions illustrate the main pillars of the Protocol: integration of water and health polices; adaptation to climate change; surveillance of water-related disease; assistance and cooperation; and involving civil society in decision-making.

Opening session

Gheorghe Constantin, General Director, General Directorate for Water Management, Romania
Roberto Menia, Under-Secretary, Ministry for the Environment, Land and Sea, Italy
Maria Neira, Director, Public Health and Environment, WHO headquarters
Francesca Bernardini, Co-Secretary of the Protocol on Water and Health, United Nations Economic Commission for Europe

Session 1. Developing integrated strategies on water and health

Chair: Pierre Studer

Advantages and challenges of setting targets under the Protocol
Pierre Studer, Swiss Federal Office of Public Health, Switzerland and Chair, Task Force on Indicators and Reporting

Setting national targets in Finland
Mikko Paunio, Ministry of Social Affairs and Health, Finland

Setting national targets in the Republic of Moldova
Ion Shalaru, National Centre for Public Health, Republic of Moldova

The relationship between the EU acquis communautaire and the Protocol
Mihály Kadar, Institute for Environmental Health, Hungary

Discussion

Session 2. The Protocol as a climate change adaptation tool: achievements of the Task Force on Extreme Weather Events

Chair: Luciana Sinisi

Water supply and sanitation in extreme weather events
Luciana Sinisi, Institute of Environmental Research and Protection, Italy and Chair, Task Force on Extreme Weather Events

The Protocol as a climate change adaptation tool
Roger Aertgeerts, Co-secretary, Protocol on Water and Health, WHO European Centre for Environment and Health, WHO Regional Office for Europe

Drinking-water-borne outbreak due to karst flooding – environmental health aspect
Gyula Dura, Institute for Environmental Health, Hungary

Climate change-induced water scarcity and adaptation strategies in the eastern Mediterranean
Manfred Lange, Energy, Environment and Water Research Centre, The Cyprus Institute

Socioeconomic benefits of adaptation policy
Jim Foster, Drinking-water Inspectorate, United Kingdom and Head, WHO Collaborating Centre on Drinking-water Safety

Multisectoral cooperation
Dominique Gatel, European Federation of National Associations of Drinking-Water Suppliers and Waste Water Services (EUREAU)

Discussion

Session 3. Surveillance of water-related diseases

Chair: Enzo Funari

Current importance of water-related diseases in Europe and emerging threats
Enzo Funari, Higher Institute of Public Health, Italy and Chair, Task Force on Water-related Disease Surveillance

Water and health in Europe
Thomas Kistemann, Professor, Institute for Hygiene, University of Bonn, Germany and Head, WHO Collaborating Centre for Health-Promoting Water Management and Risk Communication

Situation with drinking-water availability and technique for estimating health damage in Amudarya delta
Iskander Rusiev, Scientific Information Centre of the Interstate Coordination Water Commission, Uzbekistan

Small-scale water supplies in the European Region: why do they matter?
Oliver Schmoll, Federal Environmental Agency, Germany and WHO Collaborating Centre for Research on Drinking-water Hygiene

Discussion

Session 4. Water cooperation opens up new possibilities and new partners

Chair: Carola Björklund

Water cooperation opens up new possibilities and new partners
Carola Björklund, Ministry of Foreign Affairs, Norway and Chair, ad hoc Project Facilitation Mechanism

International cooperation in implementing the Protocol on Water and Health
Svitlana Nigorodova, Deputy Head, Ministers' Office, Ministry of the Environment, Ukraine

European Bank for Reconstruction and Development water investment in central Asia
Paul Covenden, Municipal and Environmental Infrastructure Team, European Bank for Reconstruction and Development (EBRD), Tbilisi

Discussion

Session 5. The Protocol on Water and Health as a tool for environment and health democracy

Chair: Charles Saout

Equitable access to water
Charles Saout, Ministry of Health, Youth, Sports and Associations, France

Ensuring compliance and implementation
Magdalena Bar, Protocol Compliance Committee, Poland

The role of multilateral organizations – how the United Nations Development Programme supports implementation of the Protocol
Jürg Staudenmann, Water Governance Adviser, United Nations Development Programme Regional Bureau for Europe and the Commonwealth of Independent States

The role of civil society in improving water and health
Sascha Gabizon, Women in Europe for a Common Future

Discussion

Closure

Media workshop

"And finally here we are!" Introductions and expectations

A virtual walk through the Conference

Cristiana Salvi and Franklin Apfel

The big show – the Conference programme
Meeting the protagonists – the Conference participants
On the side – the satellite events (seminars, workshops, standing coffees (with presentations), poster sessions)

A guided tour of the Conference premises

Building your Parma experience

Alex Kirby, Xhemal Mato, Yuri Eldyshev

Headlines – the key messages
Quotes – the key players
Images – the key moments

Making your Parma story

All moderators

Group work and discussion

Meeting the youth

Exclusive dialogue with young people about their input into the Conference outcomes

WHO Children's Environment and Health Action Plan for Europe (CEHAPE) Youth Network workshop

Presentations by Italian Environment Authority of Parma

Ecologically sustainable building where the youth event was held
Sustainable waste management by Parma Municipality (recycle, reuse and treatment, "ecological waste island")
Presentation on reforestation through bio-compensation with possible link to climate issues
Presentation of the Parma municipal project on dispersion of water in water pipelines

Preparatory activities for the Ministerial Conference

WHO young journalists network (WHY Network)

Introduction and presentation
Discussions and networking with WHO CEHAPE Youth Network

Workshop with WHO on physical activity project

Site visits

Ecological waste platform
Recycled paper factory
Reforested forest through bio-compensation
Walk to the Ducal Garden
Experimental installation to detect water losses in the pipelines

Side events (10–11 March 2010)

Symposia

The right to safe water and sanitation in schools
Women in Europe for a Common Future

Protecting and improving human health through strategic environmental assessment
International Association for Impact Assessment

Indoor air quality in Europe. Preventing and reducing respiratory diseases
Regional Environmental Center, European Commission Joint Research Centre, Ministry of Health of Italy

COPE. How children can make a difference regarding environmental and health problems
Nordic Council of Ministers

How human biomonitoring supports environment and health policy: European feasibility study on mothers and children
University of Leuven, Belgium

Environment and health inequalities
Environment Agency, United Kingdom

Transport, Health and Environment Pan-European Programme: from a policy framework to supporting action in Member States
United Nations Economic Commission for Europe, WHO Regional Office for Europe

European research on the health effects of changing environment
European Commission Directorate-General for Research

Youth participation in CEHAPE Regional Priority Goals in countries
Health and Environment Alliance

Environmental influences on children's respiratory health
European Respiratory Society

Social inequalities in occupational health
Italian National Institute for Occupational Safety and Prevention, WHO Regional Office for Europe

Preventing injuries in Europe – from international cooperation to national implementation
European Commission Directorate-General for Health and Consumers, WHO Regional Office for Europe

Safe water and healthy water in a changing environment
European Environment Agency, Ministry for the Environment, Land and Sea, Italy

Tools for climate change adaptation
European Centre for Disease Prevention and Control

Reducing the environmental impact of the food system: lessons from traditional food cultures
Hellenic Health Foundation, Food and Agriculture Organization of the United Nations, WHO Regional Office for Europe, Institute of Food Research

Implementing CEHAPE Regional Priority Goal 2: children and youth-friendly mobility management – good practice and lessons learned
Federal Ministry of Agriculture, Forestry, Environment and Water Management, Austria

Electromagnetic fields (radiofrequencies) and health: an update of expert assessment and recommendations for actions to reduce exposures and for research
French Agency for Environmental and Occupational Health Safety

Environmental noise and health: the European Network on Noise and Health
Queen Mary, University of London

Seafood safety: delivering the benefits of a healthy diet and reducing risks from environmental contamination
European Food Safety Authority, WHO Regional Office for Europe

Pilot projects on protecting health from climate change
Federal Ministry for the Environment, Nature Conservation and Nuclear Safety, Germany

European regional perspectives on environment and health
WHO Regions for Health Network

Children's health and the environment: how does current European research respond to public health priorities?
European Commission Directorate-General for Research, International Society for Environmental Epidemiology

Sustainable actions to master a changing environment
World Business Council for Sustainable Development

Food for health and sustainable growth
Ministry for the Environment, Land and Sea, Italy, Ministry of Health, Italy

Standing coffees

French actions on health and environment
Ministry of Sustainable Development, France

Research options and priorities concerning environment and health in Italy
National Research Council, Italy

Air pollution and health. EpiAir: the Italian surveillance programme on the short-term effects of air pollution
Department of Epidemiology, Lazio Region Health Service, Rome, Italy

The Italian strategic programme "Environment and health"
National Institute of Health, Italy

Housing and health
Ministry of Health, Portugal

Training of schoolchildren – toxicology in the classroom
United Nations Environment Programme

Atlas of our living environment: applying geo-ICT innovations to inform citizens and professionals about the state of their living environment
National Institute for Public Health and the Environment, the Netherlands

Environment, health, children and climate change: integrated actions between municipalities and territorial practitioners
Arezzo Healthy Cities Project and Italian Society of Doctors for the Environment, Italy

Poster sessions

The second French NEHAP
Vincent Delporte, Ministry of Sustainable Development, France

Elfe, French Birth Cohort Study
Stéphanie Vandentorren, Frédéric de Bels, Sandra Sinno-Tellier, Christophe Declercq, Adeline Floch-Barnaud and the Elfe team, Institut de veille sanitaire (InVS), France

Www.substitution-cmr.fr: a tool to support the substitution of CMR substances
Sophie Robert, Aurélie Niaudet, Céline Dubois, Pierre Lecoq, Nathalie Ruaux and members of AFSSET Working group "CMR", AFSSET (French Agency for Environmental and Occupational Health Safety), France

Short asbestos fibres: potential for exposures and health risks to the general population
Guillaume Boulanger, Amandine Paillat, and the AFSSET Working group "Short asbestos fibres and thin asbestos fibres", AFSSET (French Agency for Environmental and Occupational Health Safety), France

ERA-ENVHEALTH: coordination of national environment and health research programmes
Adrienne Pittman, AFSSET (French Agency for Environmental and Occupational Health Safety) and Mohssine El Kahloun, BELSPO (Belgian Federal Science Policy Office) on behalf of the ERA-ENVHEALTH Project partners

Development of the national strategy of biomonitoring in France
Clémence Fillol, Frédéric De Bels, Agnès Lefranc, Georges Salines, Institut de veille sanitaire (InVS), France (www.invs.sante.fr)

Health effects of chronic noise exposure in children in a changing environment
J. Horn, L. Hülsmeier, J. Fels, M. Vorländer, I. Koch, V. Lawo, W. Dott, Achen University, Germany

Housing and health
Claudia Weigert, Ministry of Health, Division of Environmental Health, Portugal

Risky conditions in tourism facilities
Department of Public Health, Medical School, Adnan Menderes University, Aydin, Turkey

Local health effect screening
Peter van den Hazel, Public Health Services Gelderland-Midden, the Netherlands

Dutch Knowledge Platform on EMF: between science and public
Ronald van der Graaf, National Institute for Public Health and the Environment (RIVM), the Netherlands

Few environmental factors are responsible for most of the environmental burden of disease
EBoDE Working Group, National Institute for Health and Welfare, Finland

Europe leads the management of chemical risks – with marginal impact on environmental burden of disease
EBoDE Working Group, National Institute for Health and Welfare, Finland

According to the EBoDE project: environmental factors contribute significantly to the burden of disease in Europe
Jurgen Buekers, Annette Prüss-Üstün and the EBoDE Working Group, National Institute for Health and Welfare, Finland

The air that children breathe: indoor air quality and health effects in elementary schools in Austria
Hans-Peter Hutter, Hanns Moshammer, Karl Kociper, Kathrin Piegler, Michael Kundi, Institute of Environmental Health, Centre for Public Health, Medical University Vienna; Peter Wallner, Medicine and Environmental Protection, Vienna, Austria; Philipp Hohenblum, Maria Uhl, Sigrid Scharf, Jürgen Schneider, Federal Environmental Agency, Vienna, Austria; Claudia Gundacker, Karl Wittmann, Department Ecotoxicology, Centre for Public Health, Medical University Vienna; Peter Tappler, Center for Architecture, Construction and Environment, Danube University Krems, Austria

Blood relatives: human biomonitoring of industrial chemicals in Austrian families
Hans-Peter Hutter, Daniela Haluza, Kathrin Piegler, Livia Borsoi, Hanns Moshammer, Peter Wallner, Michael Kundi, Institute of Environmental Health, Centre for Public Health, Medical University Vienna; Philipp Hohenblum, Sigrid Scharf, Environment Agency Austria, Vienna, Austria

Indoor environment in Belgian nurseries: from the demand to the offer
M.C. Dewolf, F.Charlet, M. Roger, Hygiène Publique in Hainaut, Mons, Belgium; M. Kuske, Service d'Analyse des Milieux Intérieures de la Province du Luxembourg, Marloie, Belgium; S. Bladt, Cellule Regionale d'Intervention en Pollution Intérieure, Brussels, Belgium; C. Chasseur, Scientific Institute of Public Health, Brussels, Belgium; A. Worobiec, B. Horemans, Chemical Department, University of Antwerp, Belgium; P. Biot, Federal Public Service Health, Food Chain Safety and Environment, Brussels, Belgium; M. Mampaey, Environment, Nature and Energy Department, Flemish Government, Brussels, Belgium; M.P. Berhin, N. Vanderheyden, Office de la Naissance et de l'Enfance, Brussels, Belgium; H. Peeters, Kind en Gezin, Brussels, Belgium; M. Verlaek, Medisch Milieukundigen van het Logo en het Vlaams Instituut voor Gezondheidspromotie en ziekte preventie, Belgium

Aphekom – The science/decision interface in air quality policy: lessons on stakeholder and citizens' expression in local participatory processes
Yorghos Remvikos, Centre d'Economie et d'Ethique pour l'Environnement et le Développement, UVSQ, France; Catherine Bouland, Brussels Institute for the Management of the Environment, IBGE, Belgium; Sylvia Medina, Institut de veille saniatire (InVS), France on behalf of the Aphekom network (www.aphekom.org)

Fight against insanitary housing: definitions and results of the French national action plan
Caroline Paul, Ministry of Health, France

Indoor air quality: a major axis of the French NEHAPs
Vincent Delporte, Ministry of Sustainable Development, France

Charters for towns that are implementing actions in relation to a healthy diet and physical activity (RPG II)
Michel Chauliac, Ministry of Health, France

Communication on regional priorities in children's environmental health
J. Linnemann, M. Otto, K.E. von Muhlendahl, Professor of Paediatrics, German Academy of Paediatrics

The Aphekom project – a literature review on air pollution interventions and their impact on public health
Susann Henschel, Focas, Dublin Institute of Technology, Ireland; Patrick Goodman, Dublin Institute of Technology, Ireland; Sylvia Medina, Institut de veille sanitaire (InVS), France on behalf of the Aphekom network (www.aphekom.org)

Disparities of childhood obesity in Italy
Angela Spinelli, Giovanni Baglio, Anna Lamberti, Alberto Perra, Gabriele Fontana, Chiara Cattaneo, National Institute of Health, Rome, Italy; Daniela Galeone, Lorenzo Spizzichino, Maria Teresa Menzano, Ministry of Health, Rome, Italy; Nancy Binkin, United Nations Children's Fund, New York, United States of America

Smoke-free moms
Daniela Galeone, Lorenzo Spizzichino, Maria Teresa Menzano, Maria Teresa Scotti, Ministry of Health, Rome, Italy; Luca Sbrogiò, Alessandra Schaivinato, Prevention Department, LHU 9 Treviso, Italy

"Forchetta e scarpetta" (fork and sneakers): programme to promote healthy lifestyles in children and adolescents
Daniela Galeone, Lorenzo Spizzichino, Maria Teresa Menzano, Maria Teresa Scotti, Ministry of Health, Rome, Italy; Maria Teresa Silani, Silvana Teti, Ministry of Education, Rome, Italy; Tiziano Fazzi, Civicamente Srl., Italy

Multimedia communication campaigns: "smoking kills; defend yourself!"
Daniela Galeone, Lorenzo Spizzichino, Maria Teresa Menzano, Alfredo D'Ari, Daniela Rodorigo, Ministry of Health, Rome, Italy

Nationwide programme for improving indoor air quality in Dutch schools
Merel Linthorst, Dutch community health services (GGD), the Netherlands

Smoke-free environments in Italy – monitoring of the Italian law to protect people from passive smoke
Daniela Galeone, Maria Teresa Menzano, Lorenzo Spizzichino, Ministry of Health, Rome, Italy

An early warning system for environmental health impacts, risk assessment methodologies and health standards for ambient air quality
Valery Filonov, Director of the Republican Scientific Practical Centre of Hygiene, Belarus; Irina Zastenskaja, Deputy Director of the Republican Scientific Practical Centre of Hygiene, Belarus; Tatyana Naumenko, Manager laboratory of a complex risk assessment of environmental factors of the Republican Scientific Practical Centre of Hygiene, Belarus

Health impacts of climate change: assessing adaptation needs for health surveillance systems in France
Mathilde Pascal, Dounia Bitar, Christophe Declercq, Loïc Josseran, Anne-Catherine Viso, Sylvia Medina, on behalf of the climate change working group, Institut de veille sanitaire (InVs), France

An example of cross-sector coordination: the fight against the spread of an invasive and very allergenic species, common ragweed
Caroline Paul, Ministry of Health, France

Protecting present and future generations – implementing lessons from the WHO Regional Office for Europe book: "Public health significance of urban pests"
Graham Jukes, Chief Executive, Jonathan Peck, member of the National Pest Advisory Panel, Chartered Institute of Environmental Health, London, United Kingdom

Health and environment capacity building
Peter van den Hazel, Public Health Services Gelderland-Midden, the Netherlands

Gaining health: the Italian strategy to prevent noncommunicable diseases
Daniela Galeone, Lorenzo Spizzichino, Maria Teresa Menzano, Ministry of Health, Rome, Italy

Annex 8. Participants

Representatives

Member States

Albania

Dr Petrit Vasili
Minister of Health

Mr Romeo Zegali
Director, European Union Integration and Foreign Relations, Ministry of Health

Andorra

Mr Jesus de Tena-Guillen
State Secretary, Ministry of Health, Welfare and Labour

Ms Margarida Coll
Director, Public Health, Ministry of Health, Welfare and Labour

Mr Xavier Cuenca
Director of Environment, Ministry of Land Management, Environment and Agriculture

Armenia

Dr Tatul Hakobyan
Deputy Minister of Health

Mr Arman Melkonyan
Adviser to the Minister of Health

Mr Viktor Martirosyan
Director, Environmental Projects Centre, Ministry of Nature Protection

Dr Anahit Aleksandryan
Head, Department of Hazardous Substances and Waste Management, Ministry of Nature Protection

Austria

Dr Reinhard Mang
Secretary-General, Federal Ministry of Agriculture, Forestry, Environment and Water Management

Mr Robert Thaler
Head, Division V/5 – Transport, Mobility, Human Settlement and Noise, Federal Ministry of Agriculture, Forestry, Environment and Water Management

Mr Günter Liebel
Director-General and Head, Department of General Environmental Policy, Federal Ministry of Agriculture, Forestry, Environment and Water Management

Dr Veronika Holzer
Deputy Head of Department, Federal Ministry of Agriculture, Forestry, Environment and Water Management

Dr Martina Reisner-Oberlehner
Expert, Division V/2, Federal Ministry of Agriculture, Forestry, Environment and Water Management

Dr Fritz Wagner
Deputy Director, Prevention and Health Promotion, Ministry of Health

Ms Cosima Pilz
Styrian Environmental Education Centre

Azerbaijan

Dr Ogtay Shiraliyev
Minister of Health

Mr Mammadhuseyn Muslumov
Director, National Environmental Monitoring Department, Ministry of Ecology and Natural Resources

Dr Samir Abdullayev
Head, International Relations Department, Ministry of Health

Belarus

Dr Robert Chasnoyt
First Deputy Minister of Health

Belgium

Mr Philippe Henry
Minister of the Environment, Spatial Planning and Mobility, Walloon Region

Mr Frédéric Chemay
Environment Adviser, Office of the Federal Minister for Climate and Energy

Mrs Laetitia Theunis
Office of the Minister of the Environment, Spatial Planning and Mobility, Walloon Region

Mr Eric Van Duyse
Press and Communications Officer, Office of the Minister of the Environment, Spatial Planning and Mobility, Walloon Region

Mr François Cornet d'Elzius
Consul-General, Milan, Italy

Ms Leen Meulenbergs
Head of Service, International Relations, Federal Public Service for Health, Food Chain Safety and Environment

Mr Pierre Biot
Attaché, Directorate-General for the Environment, Federal Public Service for Health, Food Chain Safety and Environment

Dr Yseult Navez
Coordinator, Health and Environment Desk, Federal Public Service for Health, Food Chain Safety and Environment

Dr Catherine Bouland
Head, Department of Health and Indoor Pollution, Brussels Institute for Management of the Environment

Mr Francis Brancart
Director, Environment Policy, Department for European Policies and International Agreements, Directorate-General for Agriculture, Natural Resources and Environment, Walloon Region

Ms Maja Mampaey
Policy Adviser, Environment and Health Unit, Department for Environment, Nature and Energy, Flemish Government

Ms Sofie Vanmaele
Adviser, Nature and Energy Department, International Environmental Policy Division, Department for Environment, Nature and Energy, Flemish Government

Dr Özlem Bozkurt
Environmental Health Care Worker, Division of Public Health Surveillance, Flemish Agency for Care and Health, Flemish Ministry of Welfare, Public Health and Family

Ms Emmanuèle Bourgeois
Programme Manager, Federal Science Policy Office

Bosnia and Herzegovina

Ms Mirha Ošijan
Senior Specialist, Department for Health, Ministry of Civil Affairs

Dr Senad Oprašić
Senior Specialist, Ministry of Foreign Trade and Economic Relations

Mr Emil Balavac
Commission for the Coordination of Youth Issues, Ministry of Civil Affairs (Official Youth Representative)

Bulgaria

Dr Bozhidar Nanev
Minister of Health

Professor Todorka Kostadinova
Deputy Minister of Health

Ms Emiliya Kraeva
Head, International Cooperation Department, Ministry of Environment and Water

Dr Mariana Barouh
Chief Expert in Environmental Policy, Department of Environmental Strategy and Programmes, Ministry of Environment and Water

Mr Zlati Katzarski
Head, International Humanitarian Organizations, Human Rights Directorate, Ministry of Foreign Affairs

Croatia

Dr Ante-Zvonimir Golem
State Secretary, Ministry of Health and Social Welfare

Mr Tomislav Vidošević
Ambassador of the Republic of Croatia to the Republic of Italy

Ms Sibila Zabica
Adviser for European Integration, Office of the Minister of Health and Social Welfare

Dr Krunoslav Capak
Deputy Director, Croatian Institute for Public Health

Ms Marina Prelec
Junior Advisor, Department for International Cooperation, Ministry of Environmental Protection, Physical Planning and Construction

Mrs Lidija Lukina Kezic
Consular Adviser, Consulate-General of Croatia

Cyprus

Dr Christos G. Patsalides
Minister of Health

Dr Andreas Polynikis
Chief Medical Officer, Ministry of Health

Dr Stella Michaelidou-Canna
National Committee on Environment and Children's Health

Mr George Campanellas
Administrative Officer, Office of the Minister of Health

Czech Republic

Dr Růžena Kubínová
Head, Department of Environmental Health, National Institute of Public Health

Ms Alena Marková
Head, Strategies Unit, Department of Environmental Policy, Ministry of the Environment

Denmark

Mr Steffen Egesborg Hansen
Head of Division, Ministry of the Interior and Health

Mr Henrik Søren Larsen
Head, Chemicals Division, Danish Environmental Protection Agency, Ministry of the Environment

Dr Lis Keiding
Specialized Medical Officer, Centre for Health Promotion and Disease Prevention, Ministry of the Interior and Health

Ms Mona Mejsen Westergaard
Senior Adviser, International Environmental Issues, Danish Environmental Protection Agency, Ministry of the Environment

Dr Niss Skov Nielsen
Special Adviser, Ministry of the Interior and Health

Estonia

Mr Jaanus Tamkivi
Minister of the Environment

Ms Aive Telling
Environmental Health and Chemical Safety Unit, Department of Public Health, Ministry of Social Affairs

Mrs Reet Pruul
Senior Officer, Environmental Management and Technology, Ministry of the Environment

Ms Kristina Aare
Official Youth Representative

Finland

Dr Lea Kauppi
Director-General, Finnish Environment Institute

Dr Mikko Paunio
Senior Medical Officer, Department for Promotion of Welfare and Health, Ministry of Social Affairs and Health

Ms Outi Kuivasniemi
Ministerial Adviser, Ministry of Social Affairs and Health

Ms Eija Lumme
Counsellor, Ministry of the Environment

Professor Matti Jantunen
Department of Environmental Health, National Institute of Health and Welfare

Mr Tomi Nieminen
Official Youth Representative

France

Professor Didier Houssin
Director-General for Health, Health Division, Ministry of Health and Sport

Ms Patricia Blanc
Director, Ministry of Sustainable Development

Mr Charles Saout
Deputy Director, Environment and Food Department, Health Division, Ministry of Health and Sport

Ms Géraldine Bonnin
WHO Programme Officer, Delegation for European and International Affairs, Ministry of Health and Sport

Georgia

Mr Alexander Kvitashvili
Minister of Labour, Health and Social Affairs

Professor Nikoloz Pruidze
Deputy Minister, Ministry of Labour, Health and Social Affairs

Mr George Zedginidze
Deputy Minister, Ministry of Environmental Protection and Natural Resources

Ms Nino Mirzikashvili
Head, International Relation Department, Ministry of Labour, Health and Social Affairs

Germany

Mrs Annette Widmann-Mauz
Parliamentary State Secretary, Federal Ministry of Health

Mrs Karin Knufmann-Happe
Director, Department for Prevention, Health Protection, Disease Control and Biomedicine, Federal Ministry of Health

Dr Ute Winkler
Head, Division for Basic Issues of Prevention, Self-help and Environmental Health Protection, Federal Ministry of Health

Ms Gabriela Girnau
Adviser to the Parliamentary State Secretary, Federal Ministry of Health

Dr Peter Pompe
Head, Department for Protocol, International Visitors, Relations with Domestic and Foreign Representations, Language Services, Federal Ministry of Health

Dr Stephan Böse-O'Reilly
Board Member, German Network for Children's Health and Environment

Ms Katharina Suntrup
Interpreter, Ministry for the Environment, Nature Conservation and Nuclear Safety

Mr Alexander Nies
Deputy Director-General, Federal Ministry for the Environment, Nature Conservation and Nuclear Safety

Ms Sonja Niehoff
Personal Adviser to the State Secretary, Federal Ministry for the Environment, Nature Conservation and Nuclear Safety

Dr C. Jutta Litvinovitch
Head, Division for Health Impacts of Climate Change and Environment-related Food Safety, Federal Ministry for the Environment, Nature Conservation and Nuclear Safety

Dr Birgit Wolz
Head, Division for Environment and Health, Federal Ministry for the Environment, Nature Conservation and Nuclear Safety

Dr Björn Ingendahl
Division for Environment, Health and Consumer Protection, Federal Ministry for the Environment, Nature Conservation and Nuclear Safety

Dr Hedi Schreiber
Head, Health Effects Assessment, Division for Environment, Hygiene and Medicine, Federal Environment Agency

Dr Marike Kolossa-Gehring
Head of Section, Division for Toxicology and Health-related Environmental Monitoring, Federal Environment Agency

Greece

Ms Vassiliki Karaouli
Director, Sanitary Engineering and Environmental Health, Ministry of Health

Dr Athena Mourmouris
Head, Department of Geographical Information Systems and Observatory for Physical Planning, Ministry of Environment, Energy and Climate Change

Hungary

Dr Melinda Medgyaszai
Ministerial Commissioner for International Affairs, Ministry of Health

Dr Tibor Farago
State Secretary for Environment and Climate Policy, Ministry of Environment and Water

Dr Gyula Dura
Director, Institute of Environmental Health

Dr Anna Margit Paldy
Deputy Director-General, Institute of Environmental Health

Dr Balint Dobi
Head, Department for Environmental Conservation, Ministry of Environment and Water

Dr Zsuzsanna Pocsai
Chief Adviser, Ministry of Environment and Water

Dr Zsuzsanna Tomka
Adviser, Ministry of Health

Ireland

Ms Siobhan McEvoy
Chief Environmental Health Officer, Environmental Health Unit, Department of Health and Children

Israel

Mr Yaakov Litzman
Deputy Minister of Health

Dr Itamar Grotto
Director, Public Health Services, Ministry of Health

Mr Shalom Goldberger
Chief Engineer, Environmental Health, Ministry of Health

Dr Orna Matzner
Head, Science Unit, Office of the Chief Scientist, Ministry of Environmental Protection

Ms Beth-Eden Kite
Director of Training, Centre for International Cooperation, Ministry of Foreign Affairs

Italy

Professor Ferruccio Fazio
Minister of Health

Ms Stefania Prestigiacomo
Minister of Environment, Land and Sea

Dr Corrado Clini
Director-General, Department for Sustainable Development, Climate Change and Energy, Ministry of Environment, Land and Sea

Dr Fabrizio Oleari
Director, Directorate-General for Prevention, Ministry of Health

Mr Roberto Menia
Undersecretary of State, Ministry of Environment, Land and Sea

Mr Antonio Bernardini
Diplomatic Counsellor, Ministry of Environment, Land and Sea

Ms Paola Lucarelli
Deputy Head, Office of the Minister, Ministry of Environment, Land and Sea

Mr Salvatore Bianca
Head, Press Office, Ministry of Environment, Land and Sea

Ms Manuela Campisi
Head, Technical Secretariat of the Minister, Ministry of Environment, Land and Sea

Mr Fabrizio Penna
Technical Secretariat of the Undersecretary of State, Ministry of Environment, Land and Sea

Mr Luigi Pulvirenti
Press Office, Ministry of Environment, Land and Sea

Ms Simona Di Cresce
Technical Secretariat of the Undersecretary of State, Ministry of Environment, Land and Sea

Mrs Giuliana Gasparrini
Head of Division and National Focal Point, Department for Sustainable Development, Climate Change and Energy, Ministry of Environment, Land and Sea

Ms Martina Hauser
Balkans Task Force, Department for Sustainable Development, Climate Change and Energy, Ministry of Environment, Land and Sea

Mr Massimo Cozzone
Senior Officer, Department for Sustainable Development, Climate Change and Energy, Ministry of Environment, Land and Sea

Ms Benedetta Dell'Anno
Policy Adviser, Department for Sustainable Development, Climate Change and Energy, Ministry of Environment, Land and Sea

Mr Alessandro Negrin
Expert, Department for Sustainable Development, Climate Change and Energy, Ministry of Environment, Land and Sea

Mr Cristiano Piacente
Expert, Department for Sustainable Development, Climate Change and Energy, Ministry of Environment, Land and Sea

Dr Alessandra Burali
Expert, Department for Sustainable Development, Climate Change and Energy, Ministry of Environment, Land and Sea

Mr Vincenzo Grimaldi
Commissioner, Higher Institute for Environmental Protection and Research

Mr Emilio Santori
Subcommissioner, Higher Institute for Environmental Protection and Research

Dr Luciana Sinisi
Unit Head, Environmental Determinants of Health, Higher Institute for Environmental Protection and Research

Mr Mario Alberto di Nezza
Head of Cabinet, Ministry of Health

Mr Manuel Jacoangeli
Diplomatic Counsellor, Ministry of Health

Dr Francesca Basilico
Head, Technical Secretariat of the Minister, Ministry of Health

Dr Adelmo Grimaldi
Head, Secretariat of the Minister, Ministry of Health

Dr Romano Marabelli
Head, Disease Prevention/Communication, Welfare, Health and Social Affairs, Ministry of Health

Dr Francesco Cicogna
Senior Medical Officer, Directorate-General for the European Union and International Relations, Ministry of Health

Dr Liliana La Sala
Director, Health and Environment Office, Directorate-General for Prevention, Ministry of Health

Dr Daniela Galeone
Director, Office II, Department of Prevention and Communication, Ministry of Health

Dr Pier Giuseppe Facelli
Senior Veterinary Officer, Department of Veterinary Public Health, Nutrition and Food Safety, Ministry of Health

Dr Annamaria De Martino
Medical Officer, Directorate-General for Prevention, Ministry of Health

Dr Annunziatella Gasparini
Director, Minister's Press Office, Ministry of Health

Dr Loredana Di Leginio
Minister's Press Officer, Ministry of Health

Kazakhstan

Dr Kenes Ospanov
Head, Committee for National Sanitary/Epidemiological Control, Ministry of Health

Ms Umitzhan Itekbayeva
International Expert, Kazakh Ecology and Climate Research Institute, Ministry of Environmental Protection

Miss Gulaiym Tnymbergen
Expert, International Cooperation Division, Ministry of Health

Kyrgyzstan

Dr Marat Mambetov
Minister of Health

Mr Arstanbek Davletkeldiev
Director, State Agency for Environmental Protection and Forestry

Dr Ainash Akynovna Sharshenova
Head, Department of Environmental Health, Scientific and Production Centre for Preventive Medicine

Latvia

Ms Astra Kurme
Ambassador of the Republic of Latvia to the Republic of Italy

Lithuania

Mr Audrius Šceponavičius
Director, Public Health Department, Ministry of Health

Luxembourg

Dr Yolande Wagener
Director, Division of Preventive and Social Medicine, Health Directorate, Ministry of Health

Mr Ralph Baden
Materials Engineer, Division of Occupational Health, Health Directorate, Ministry of Health

Mr Marc Fischer
Engineer, Ministry of Health

Malta

Dr Joseph Cassar
Minister for Health, the Elderly and Community Care

Mr Malcolm Vella Haber
Private Secretary to the Minister for Health, the Elderly and Community Care

Dr Ray Busuttil
Director-General (Health), Public Health Regulation Division, Ministry for Health, the Elderly and Community Care

Mr John Attard Kingswell
Director, Environmental Health, Public Health Regulation Division, Ministry for Health, the Elderly and Community Care

Mr Franck Lauwers
Senior Environment Protection Officer, European Union and Multilateral Affairs Unit, Environment and Planning Authority

Monaco

Dr Anne Negre
Director, Directorate for Health and Social Work, Department of Social Affairs and Health

Mr Frederic Pardo
Head, External Relations, Department of External Relations, Directorate of International Affairs, Ministry of State

Montenegro

Professor Dr Miodrag Radunovic
Minister of Health

Dr Rajko Strahinja
Assistant Minister, Ministry of Health

Ms Marina Miskovic
Senior Adviser, Department for Nature Protection and Environmental Assessment, Ministry of Spatial Planning and the Environment

Netherlands

Mr Hugo G. von Meijenfeldt
Director, International Affairs/Climate Envoy, Ministry of Housing, Spatial Planning and the Environment

Dr Julie Ng-A-Tham
Coordinator, Environment and Health, Ministry of Housing, Spatial Planning and the Environment (EEHC member)

Mr Tom van Teunenbroek
Nanotechnologies Specialist, Environment and Health, Ministry of Housing, Spatial Planning and the Environment

Mr Paul Huijts
Director-General for Public Health, Ministry of Health, Welfare and Sport

Mr Fred Lafeber
Head, Global Affairs Unit, Ministry of Health, Welfare and Sport

Ms Michaela Hogenboom
Commission on Sustainable Development, Ministry of Health, Welfare and Sport (Official Youth Representative)

Dr Marc Sprenger
Director-General, National Institute for Public Health and the Environment, Ministry of Health, Welfare and Sport

Norway

Ms Vigdis Roenning
Senior Adviser, Department of Public Health, Ministry of Health and Care Services

Ms Hilde Moe
Senior Adviser, Department of Regional Planning, Ministry of the Environment

Mr Kjetil Tveitan
Assistant Director-General, Ministry of Health and Care Services

Dr Jon Hilmar Iversen
Deputy Director, Department for Primary Health Care Services, Norwegian Health Directorate (EEHC Chair)

Ms Bente Elisabeth Moe
Senior Adviser, Department for Primary Health Care Services, Norwegian Health Directorate

Ms Helene Kaltenborn
Official Youth Representative

Poland

Ms Ewa Kopacz
Minister of Health

Mr Krzysztof Suszek
Director, Press and Promotion, Ministry of Health

Mr Slawomir Wieslawski
Interpreter, Ministry of Health

Mr Artur Jerzy Badyda
Adviser, Political Office of the Minister of the Environment

Professor Wojciech Hanke
Professor of Environmental Epidemiology, Nofer Institute of Occupational Medicine

Dr Anna Starzewska-Sikorska
Senior Scientist, Institute for Ecology of Industrial Areas

Portugal

Mr Humberto Rosa
Secretary of State for the Environment, Ministry for Environment and Spatial Planning

Professor Maria do Céu Machado
High Commissioner of Health, Ministry of Health

Professor António Gonçalves Henriques
Director-General, Portuguese Environment Agency

Ms Patricia Veloso
Advisor to the Secretary of State, Ministry for Environment and Spatial Planning

Dr Ana Cristina Janela Bastos
Adviser, Health Department, Office of the High Commissioner for Health

Ms Claudia Weigert
Architect, Division of Environmental Health, Directorate-General for Health, Ministry of Health

Dr Regina Maria Madail Vilão
Director, Department for Environmental Policies and Strategies, Ministry for Environment and Spatial Planning

Ms Sandra Moreira
Desk Officer, Department of Environmental Policies and Strategies, Portuguese Environmental Agency, Ministry for Environment and Spatial Planning

Republic of Moldova

Professor Vladimir Hotineanu
Minister of Health

Mr Gheorghe Salaru
Minister of Environment

Professor Ion Bahnarel
Senior Scientist, Department of Public Health, National Research Centre for Preventive Medicine, Ministry of Health (EEHC member)

Mrs Evghenia Verlan
Deputy Head of Division, Ministry of Ecology and Natural Resources

Romania

Professor Adrian Streinu-Cercel
Secretary of State, Ministry of Health

Dr Maria-Mihaela Armanu
Counsellor for European Affairs, Directorate of Public Health, Ministry of Health

Dr Maria Alexandra Cucu
Director, National Centre for Health Assessment and Promotion, National Institute of Public Health

Mr Gheorghe Constantin
Director-General, Directorate-General for Water Management, Ministry of Environment and Forests

Russian Federation

Dr Marina Shevyreva
Director, Department of Health Protection and Sanitary/Epidemiological Well-being, Ministry of Health and Social Development

Mr Oleg Shamanov
Head, Division of Global Environment and Public Health, Department of International Organizations, Ministry of Foreign Affairs

Mr Viktor Baldin
Assistant to the Minister of Health and Social Development

Dr Natalia Kostenko
Head of Unit, Department of Health Protection and Sanitary/Epidemiological Well-being, Ministry of Health and Social Development

Dr Alexey Kulikov
Chief Specialist, Department of International Cooperation, Ministry of Health and Social Development

Dr Evgeny Kovalevskiy
Scientist, Research Institute of Occupational Health of Russian Academy of Medical Sciences

Dr Andrey Guskov
Deputy Chief, Sanitary Inspection, Federal Service for Surveillance of Protection of Consumer Rights and Human Well-being

San Marino

Dr Andrea Gualtieri
Director, Public Health Authority, Ministry of Health

Dr Omar Raimondi
Manager, Environment Protection Office

Serbia

Professor Dr Tomica Milosavljević
Minister of Health

Professor Dr Ivica Radović
State Secretary, Ministry for Environment and Spatial Planning

Dr Elizabet Paunović
Assistant Minister of Health for International Cooperation, Ministry of Health

Ms Biljana Filipović
Advisor for International Cooperation, Department for International Cooperation and European Integration, Ministry of Environment and Spatial Planning

Dr Tanja Knežević
Director, Dr Milan Jovanovic Batut National Institute of Public Health

Slovakia

Dr Ivan Rovny
Chief Public Health Officer

Ms Katarina Halzlova
Head, Department of Environment and Health, Public Health Authority

Dr Jan Janiga
Senior Adviser, Environmental Risk Assessment, Ministry of the Environment

Slovenia

Dr Ivan Eržen
State Secretary, Ministry of Health

Ms Marta Ciraj
Secretary, Ministry of Health

Spain

Dr Fernando Carreras Vaquer
Deputy Director-General, Environmental and Occupational Health, Ministry of Health and Social Policy

Ms Paz Valiente-Calvo
Assistant Director-General, Adaptation and Impact, Directorate-General for Environmental Quality and Assessment, State Secretariat for Climate Change, Ministry of the Environment

Dr Margarita Alonso Capitán
Technical Adviser on Environmental Health, Ministry of Health and Social Policy

Ms Ana Fresno Ruiz
Assistant Deputy Director-General, Air Quality and Industrial Environment Quality and Assessment, Ministry of the Environment

Dr Argelia Castaño
Head, Environmental Toxicology, National Centre for Environmental Health, Carlos III Health Institute

Sweden

Ms Charlotta Broman
Deputy Director, Division for Eco-Management and Chemicals, Ministry of the Environment

Mr Bo Pettersson
Senior Adviser, Public Health Policy, Ministry of Health and Social Affairs

Mr Urban Boije Af Gennas
Senior Adviser, Ministry of Health and Social Affairs

Dr Margareta Palmquist
Senior Programme Officer, Environmental and Public Health, National Board of Health and Welfare

Ms Ida Karkiainen
Official Youth Representative

Switzerland

Dr Gaudenz Silberschmidt
Head, Division of International Affairs and Vice-Director, Federal Office of Public Health

Ms Ursula Ulrich-Vögtlin
Head of Division, Multisectoral Projects, Federal Office of Public Health

Ms Aglaja Schinzel
Scientific Advisor, Political Division, Federal Department of Foreign Affairs

Ms Olivia Heller
Intern, European Child Safety Alliance (Official Youth Representative)

Tajikistan

Dr Nusratullo Salimov
Minister of Health

Mr Khursandkul Zikirov
Chairman, Environment Committee

Dr Samardin P. Aliev
Head, State Sanitary/Epidemiological Surveillance Service, Ministry of Health

Mr Firuz Nazarov
Official Youth Representative

The former Yugoslav Republic of Macedonia

Dr Bujar Osmani
Minister of Health

Professor Dragan Gjorgjev
Head of Department, Institute for Public Health, Ministry of Health

Ms Slobodanka Temova
Head of Unit, Ministry of Health

Mr Rijad Alimi
Director, Special Hospital for Children's Lung Diseases

Turkey

Dr Fehmi Aydinli
Deputy General Director, Department of Primary Health Care, Ministry of Health

Professor Çağatay Güler
Department of Public Health, Medical Faculty, Hacettepe University

Dr Aydin Yildirim
Deputy General Director, Environment Management, Department of Foreign and EU Relations, Ministry of Environment and Forestry

Turkmenistan

Mrs Shirin Rejepova
Chief Specialist, Sanitation Department, State Sanitary/Epidemiological Service, Ministry of Health and Medical Industry

Ukraine

Dr Vasyl Kniazevych
Minister of Health

Mr Taras Trotskyi
Head, Department for International Cooperation and European Integration, Ministry of Environmental Protection

Ms Zhanna Tsenilova
Head, International Department, Ministry of Health

Ms Irina Vsevolodovna Iarema
Chief Specialist and National Focal Point for the Protocol on Water and Health, Department for International Cooperation and European Integration, Ministry of Environmental Protection

Ms Anastasiya Pozikhaylo
Member, Women and Children of Ukraine (NGO), Kyiv (Official Youth Representative)

United Kingdom

Professor David Harper
Director-General and Chief Scientist, Health Improvement and Protection, Department of Health

Dr Arwyn Davies
Head, Chemicals and Nanotechnologies, Department for Environment, Food and Rural Affairs

Dr Louise Newport
Scientific Policy Manager, Health Protection, Legislation and Environmental Hazards, Department of Health

Ms Agatha Ferrão
Science Policy Coordinator, Department of Health

Mr Kyle Worgan
Official Youth Representative

Intergovernmental bodies and international organizations[1]

European Commission

Mr John Dalli
European Commissioner for Health and Consumer Policy

Ms Paola Testori Coggi
Deputy Director-General, Directorate-General for Health and Consumers

Dr Andrzej Rys
Director, Public Health and Risk Assessment, Directorate-General for Health and Consumers

Dr Laurent Bontoux
Programme Officer, Directorate-General for Health and Consumers

Mr Kevin McCarthy
Head of Sector, Public Health Research, Health Directorate, Directorate-General for Research

Mr Giulio Gallo
Administrator, Directorate-General for Health and Consumers

Ms Natacha Grenier
Administrator/Policy Officer, Healthy Environments and Injury Prevention, Directorate-General for Health and Consumers

Mr Michael Hübel
Head of Unit, Public Health and Risk Assessment, Directorate-General for Health and Consumers

Mr Harald Kandolf
Office of the European Commissioner for Health and Consumer Policy

Dr Tuomo Karjalainen
Scientific Officer, Environment Directorate, Directorate-General for Research

Dr Stylianos Kephalopoulos
Policy Support Action Leader for Health and Environment, Joint Research Centre

Ms Marina Koussathana
Directorate-General for Health and Consumers

Dr Dimitrios Kotzias
Unit Head, Joint Research Centre

Ms Elisabeth Lipiatou
Head, Climate Change and Environmental Risks, Directorate-General for Research

Dr Josefa Barrero Moreno
Administrator and Competence Group Leader, Joint Research Centre

Dr Peter Pärt
Adviser, Health and Environment Interactions, Joint Research Centre

Ms Birgit van Tongelen
Policy Officer, Biotechnology, Pesticides and Health, Directorate-General for the Environment

Dr Tomas Turecki
Project Officer, Directorate-General for Research

Ms Josépha Wonner
Assistant, Directorate-General for Health and Consumers

Mr Frank Zammit
Assistant to the European Commissioner for Health and Consumer Policy

Ms Anthia Ann Zammit
Official Youth Representative

Ms Alma Ildikó Almasi
Official Youth Representative

European Centre for Disease Prevention and Control

Professor Karl Ekdahl
Acting Director

Professor Jan Semenza
Section Head, Future Threats and Determinants, Unit of Scientific Advice

Mrs Kathryn Henriksson
Information Officer

Professor Johan Giesecke
Chief Scientist and Head of Unit, Scientific Advice, Infectious Disease Epidemiology

European Environment Agency

Professor Jacqueline McGlade
Executive Director

Dr David Stanners
Head, International Cooperation Unit

---

[1] Including accompanying advisers.

Mr André Jol
Head of Group, Vulnerability and Adaption

Dr Dorota Jarosinska
Environment and Health Project Manager, Integrated Environmental Assessments

Ms Elisabetta Scialanca
Environment and Health Project Manager

Dr Flavio Fergnani
Web and Multimedia Project Manager
European Food Safety Authority

Dr Catherine Geslain-Lanéelle
Executive Director

Dr Hubert Deluyker
Director, Scientific Cooperation and Assistance

Ms Victoria Villamar
Assistant to the Executive Director

Mr Dirk Detken
Head, Legal and Policy Affairs Unit

Professor Diána Bánáti
Chair, Management Board

Food and Agriculture Organization of the United Nations

Dr Ute Ruth Charrondière
Nutrition Officer, Nutrition Planning, Assessment and Evaluation Service

Dr Florence Egal
Nutrition Officer, Nutrition and Consumer Protection Division

Organisation for Economic Co-operation and Development

Dr Robert Visser
Deputy Director, Environment Directorate (EEHC member)

Regional Environmental Center for Central and Eastern Europe

Ms Marta Szigeti Bonifert
Executive Director (EEHC member)

Mr Zsolt Bauer
Communications

Dr Eszter Reka Mogyorosy
Expert, Business Developer

Ms Stefania Romano
Head, Italian Trust Fund

Dr Janos Zlinszky
Senior Adviser to the Executive Director (EEHC alternate)

Ms Dorottya Mogyorosi
Expert

Dr Eva Csobod
Environment and Health Topic Leader and Director, Hungary Country Office

United Nations Development Programme

Ms Kori Udovički
Assistant Administrator and Regional Director for Europe and the Commonwealth of Independent States

United Nations Economic Commission for Europe

Mr Ján Kubiš
Executive Secretary

Ms Christina von Schweinichen
Deputy Director, Environment, Housing and Land Management (EEHC Member)

Ms Ella Behlyarova
Environmental Affairs Officer, Environment, Housing and Land Management Division

Ms Francesca Bernardini
Co-Secretary, Protocol on Water and Health

Mr Tomasz Juszczak
Secretariat, Protocol on Water and Health

Mr Nicholas Bonvoisin
Environmental Affairs Officer

United Nations Environment Programme

Mr Christophe Bouvier
Director and Regional Representative for Europe

United Nations Framework Convention on Climate Change

Ms Wanna Tanunchaiwatana
Acting Coordinator, Adaptation, Technology and Science Programme

Ms Tiffany Hodgson
Assistant Programme Officer, Adaptation, Technology and Science Programme

United Nations Children's Fund

Dr Octavian Bivol
Regional Adviser, Health Systems and Policy

Ms Vilma Qahoush Tyler
Nutrition Specialist, Health and Nutrition

World Health Organization

WHO headquarters

Dr Anarfi Asamoa-Baah
Deputy Director-General

Ms Egle Granziera
Legal Officer

Dr Maria Neira
Director, Public Health and Environment

Dr Roberto Bertollini
Coordinator, Evidence and Policy

WHO Regional Office for Africa

Dr Lucien Manga
Programme Manager, Division of Prevention and Control of Communicable Diseases

WHO Regional Office for Europe

Ms Zsuzsanna Jakab
WHO Regional Director for Europe

Dr Nedret Emiroglu
Acting Director, Division of Health Programmes

Dr Enis Barış
Director, Division of Country Health Systems

Dr Francois Decaillet
Head, Brussels office

Mr Imre Hollo
Director, Division of Administration and Finance

Dr Hans Kluge
Unit Head, Division of Country Health Systems

Dr Michal Krzyzanowski
Acting Head, Bonn office

Mr Joe Kutzin
Regional Adviser, Barcelona office

Dr Lucianne Licari
Adviser, ECDC/WHO Relations, Office of the Regional Director

Dr Srdan Matic
Head, Noncommunicable Diseases and Environment, and Conference Coordinator

Dr Jose Martin Moreno
Senior Adviser

Mr Arun Nanda
Adviser

Ms Francesca Racioppi
Acting Head, Rome office

Dr Erio Ziglio
Head, Venice office

Dr Dafina Dalbokova
Consultant, Bonn Office

Mr Joris Auert
Legal Officer

Nongovernmental organizations

Eco-Forum

Ms Sascha Gabizon
Executive Director, Women in Europe for a Common Future and European Eco-Forum

Ms Demi Theodori
Coordinator, Chemicals and Health, Women in Europe for a Common Future

Ms Alexandra Caterbow
Policy Officer, Chemicals and Health, Women in Europe for a Common Future

Health and Environment Alliance

Ms Génon K. Jensen
Executive Director

Ms Joanne Vincenten
Director, EuroSafe/European Child Safety Alliance

Mr Andre Cicolella
President, Réseau Environnement Santé

Dr Hanns Moshammer
Environmental Hygienist, International Doctors for the Environment

Professor Dominique Belpomme
President, Association for Research and Treatments Against Cancer

Mr Peter van den Hazel
Section President, Environmental-related Diseases, European Public Health Association

International Trade Union Confederation

Mr Bjørn Erikson
Head, Working Environment Department

World Business Council for Sustainable Development

Dr Gernot Klotz
Executive Director, Research and Innovation, European Chemical Industries Council

Ms Loredana Ghinea
Manager, Emerging Science/Policy Issues, Research and Innovation, European Chemical Industries Council

Mr Willy De Backer
Editor, EurActiv.com

Ms Annie Mutamba
Communications Adviser, Research and Innovation, European Chemical Industries Council

Ms Carolina Susin
Adviser on Emerging Science Policy, Research and Innovation, European Chemical Industries Council

Dr Corinna Weinz
Manager, Environment and Health Concepts, Bayer AG Corporate Center for Environment and Sustainability

Ms Csilla Magyar Seinecke
Director, European Union Trade and Chemicals Policy, Dow Europe GmbH

Guest speakers

Professor Giovanni Berlinguer
Professor of Occupational Health, University of Rome, Italy

Ms Deborah Cohen
Features and Debates Editor, *British Medical Journal*, United Kingdom

Sir Andy Haines
Dean, London School of Hygiene and Tropical Medicine, United Kingdom

Dr Göran Henriksson
Senior Public Health Adviser, Västra Götaland Region, Sweden

Dr Mihály Kökény
Chairman, Parliamentary Health Committee, Hungary

Professor Sir Michael Marmot
Head, Department of Epidemiology and Public Health, University College London, United Kingdom

Professor George Morris
Consultant in Ecological Public Health, Health Protection Scotland, United Kingdom

Dr Antonio Garcia Navarro
Director-General, Carlos III Health Institute, Spain

Observers

Member States

Austria

Ms Maria Hawle
Klimabündnis Österreich

Ms Emily Hensel
Secondary School, Gaweinstal

Dr Hans-Peter Hutter
Medical Doctor, Scientist, International Society of Doctors for the Environment, Austrian Section

Professor Elisabeth Lindner
Private Technical High School, Volders

Ms Claudia Kinzl
Chief Executive Officer, Jugend-Umwelt-Netzwerk

Ms Renate Nagy
Public Officer, Federal Ministry of Agriculture, Forestry, Environment and Water Management

Ms Gudrun Redl
Jugend-Umwelt-Netzwerk

Mr Christopher Robosch
Project Manager, Offene Jugendarbeit Dornbirn

Belgium

Dr Louis Bloemen
Director, Enviromental Health Services International

Dr Ludwine Casteleyn
Coordinator, COPHES project, Université catholique de Louvain

Mrs Marie-Christine DeWolf
Project Leader, Mapping and Risk Assessment, Hainaut Vigilance Sanitaire

Dr Mohssine El Kahloun
Attaché, Belgian Federal Science Policy Office

Mr Claude Lauvaux

Dr An van Nieuwenhuyse
Programme Leader, Environmental Health Unit, Scientific Institute of Public Health

Ms Saskia Pintens
Ovio-Crioc

Dr Roel Smolders
Environment and Health Expert, Environmental Risks and Health, Vision on Technology

Mr Koen Wijnants
Environmental Health Specialist, Logo Kempen

Ms Valérie Xhonneux
Chargée de Mission, Inter-Environnement-Wallonie

Ms Dominique Mestdag-Baiwir
Event Coordinator, Ligaris Europe

Croatia

Mr Tomislav Mareelic
Driver, Embassy of the Republic of Croatia in the Republic of Italy

Czech Republic

Ms Anja Leetz
Executive Director, Health Care without Harm Europe (HCWHE)

Dr Josef Richter
Regional Institute of Public Health

Mrs Stanislava Richterova
Head, Department of Research and International Cooperation, State Health Institute

Mr Petr Severa
Head, Department of Health and Social Affairs, Regional Authority of the Usti Region

Denmark

Dr Anna-Maria Andersson
Research Director, Department of Growth and Reproduction, Rigshospitalet

Ms Francesca Viliani

Finland

Dr Otto Hänninen
Department of Environmental Health, National Institute for Health and Welfare

Ms Suvi Anneli Lehtinen
Chief, International Affairs, Finnish Institute of Occupational Health

France

Dr Séverine Deguen
Department of Occupational and Environmental Health, EHESP School of Public Health

Ms Soleane Duplan
Réseau Environnement Santé

Ms Salma Elreedy
Head, European and International Relations Unit, French Agency for Environmental and Occupational Health Safety

Mr Martin Guespereau
Director-General, French Agency for Environmental and Occupational Health Safety

Dr Renaud Lancelot
Epidemiologist, Biological Systems Department, Centre for International Cooperation on Agricultural Research for Development

Dr Sylvia Medina
Coordinator, European and International Activities, Environment and Health Department, Institut de veille sanitaire

Mr Olivier Merckel
Head of Unit, Physical Agents, New Technologies and Large Infrastructures, French Agency for Environmental and Occupational Health Safety

Ms Caroline Paul
Head Manager, Chemicals and External Environment Department, Ministry of Health and Sport

Dr Georges Salines
Department Head, Institut de veille sanitaire

Ms Marie-Alice Telle-Lamberton
Deputy Head of Department, Expertise in Environmental and Occupational Health, French Agency for Environmental and Occupational Health Safety

Dr Anne-Catherine Viso
European Topics, Scientific Directorate, Institut de veille sanitaire

Professor Denis Zmirou-Navier
Research Unit, EHESP School of Public Health

Georgia

Dr Manana Devidze
Director, Caucasus Environment

Dr Manana Juruli
Senior Researcher, Department of Toxicology, N. Makhviladze Institute of Labour, Medicine and Ecology

Professor Givi Katsitadze
Georgian Association of Toxicologists

Ms Ketevan Kiria
International Coordinator, The Greens Movement of Georgia/Friends of the Earth Georgia

Ms Rusudan Simonidze
Co-Chair, The Greens Movement of Georgia/Friends of the Earth Georgia

Germany

Dr Wolfgang Babisch
Senior Research Officer, Department of Environmental Hygiene, Division of Environment and Health, Federal Environment Agency

Dr Gabriele Bolte
Department of Environmental Health, Bavarian Health and Food Safety Authority

Professor Rainer Fehr
Head, Prevention and Innovation Department, North Rhine-Westphalia Institute of Health and Work

Dr Jeanette Miriam Horn
Institute for Hygiene and Environmental Medicine, Aachen University

Dr Reinhard Joas
Managing Director, BiPRO GmbH

Professor Thomas Kistemann
Head, WHO Collaborating Centre for Health-Promoting Water Management and Risk Communication, University of Bonn

Ms Judith Linnemann
Health Communication, German Academy of Pediatrics

Dr Doreen McBride
Academic Research Fellow, Institute for Social Medicine, Epidemiology and Health Economics, Charité – Universitätsmedizin Berlin

Dr Peter Ohnsorge
Managing Chairman, European Academy for Environmental Medicine

Dr Matthias Otto
Head of Department, Children's Environment and Health, German Academy of Pediatrics

Mr Thilo Panzerbieter
Executive Director, German Toilet Organization

Dr Alexandra Polcher
Project Manager, BiPRO GmbH

Ms Marianne Rappolder
Scientist, Federal Environment Agency

Mr Oliver Schmoll
WHO Collaborating Centre for Research and Drinking-Water Hygiene, Federal Environment Agency

Greece

Ms Carla Baer Manolopoulou
President, Clean up Greece

Dr Maria Botsivali
National Hellenic Research Foundation

Ms Fotini Kalpakioti
Youth Project Officer, Clean up Greece

Professor Antonia Trichopoulou
Vice-President, Hellenic Health Foundation

Ms Effie Vasilopoulou
Hygiene, Epidemiology and Medical Statistics, National and Kapodistrian University of Athens

Hungary

Dr Marianna Csedrekine Penzes
Health Faculty, Debrecen University

Dr Peter Rudnai
Head of Division, Environmental Health Impact Assessment, National Institute of Environmental Health

Ireland

Ms Susann Henschel
Postgraduate Research Student, Focas Institute, Dublin Institute of Technology

Italy

Ms Christina Alloti

Dr Paola Angelini
Public Health Service, Emilia-Romagna Region

Dr Massimo Aquili
Director, Office V, Directorate-General for Communication, Ministry of Health

Mr Fabio Arcuri
Project Officer, LifeGate

Dr Simona Arletti
Councillor for Environment Policies, Modena Municipality

Ms Leone Arsenio
Head, Department of Metabolic Diseases and Diabetology, Parma University Teaching Hospital

Ms Cecilia Azzali
Parma Incoming Convention Bureau

Professor Cesare Azzali
Director, Parma Industrial Union

Dr Antonella Bachiorri
Resarcher, Centro Etica Ambientale

Ms Patrizia Ballardini
Adviser, Trento Development Agency

Dr Alessandro Barchielli
Director, Epidemiology Unit, Local Health Authority, Florence

Ms Maria Chiara Barilla

Professor Giancarlo Belluzzi
Director of Office, Ministry of Health

Professor Gianfranco Beltrami
Sports medicine, Parma Province

Mr Mauro Bertoli
Technical Director, Territorial Operating Company, Enia

Dr Fabrizio Bianchi
Director of Research, Environmental Epidemiology Unit, Institute of Clinical Physiology, National Research Council

Dr Stefania Bichi
Administrative Officer, Directorate-General for Prevention, Ministry of Health

Dr Marco Biocca
Health and Social Agency, Emilia-Romagna Region

Dr Fabio Boccuni
Researcher, National Institute for Occupational Safety and Prevention

Mr Lorenzo Bono
Consultant, Ambiente Italia

Ms Gennero Cristiane Borriello

Ms Filomena Bugliaro
Federasma

Mr Ennio Cadum
Director, Department of Epidemiology and Environmental Health, Prevention and Environment Agency, Piedmont Region

Mr Paolo Caggiati
President, Parma Energy Agency

Dr Nando Campanella
Director, International Cooperation, United Hospitals of Ancona and Health Department, Marche Region

Mr Guido Canali
Architect, Parma Province

Dr Sonia Maria Margherita Cantoni
General Manager, Environmental Protection Agency, Tuscany Region

Professor Paolo Carrer
Department of Occupational and Environmental Health, Luigi Sacco University Hospital, Milan

Mr Giancarlo Castellani
Environmental Assessor, Parma Province

Professor Giovanni Cavagni
Paediatric Allergology, Bambino Gesù Pediatric Hospital, Rome

Mr Salvatore Cerracchio
Security Officer, Ministry of Health

Mr Stefano Ciafani
Scientist, National Secretariat, Legambiente (Italian League for the Environment)

Ms Eleonora Ciampini
Administrative Officer, Istituto Poligrafico e Zecca dello Stato

Dr Ferdinando Cigala
Director, Service for Prevention and Safety of the Working Environment

Mrs Federica Cingolani
Communication Agency, Rome

Dr Paolo Conti
Assistant to Professor Giovanni Berlinguer, University of Rome

Mr Stefano Coltellaci

Dr Pietro Comba
Department of Environment and Primary Prevention, Istituto Superiore di Sanità

Dr Liliana Cori
Researcher, Environmental Epidemiology Unit, Institute of Clinical Physiology, National Research Council

Mr Pierluigi Coruzzi
Director, Parma Energy Agency

Dr Emilio Cosentino
Health Officer, Ministry of Health

Mr Marco Cremonini
D'Appolonia/CETMA

Mr Giuseppe Dallara
Director, Regional Agency for Prevention and the Environment, Parma Province

Mr Tiberio D'Aloia
President, Medical Association, Parma Province

Dr Gennaro D'Amato
Director, Division of Respiratory and Allergic Diseases, Department of Chest Diseases, Antonio Cardarelli Hospital, Naples

Ms Francesca Di Maio
Institute for Environmental Protection and Research

Dr Pasquale (Lino) Di Mattia
Centre for Training and Research in Public Health

Dr Mauro Dionisio
Senior Medical Officer, Directorate-General for Prevention, Ministry of Health

Dr Dounia Ettaib
Health Assessment Unit, Milan

Mr Michele Faberi
Environment and Energy Engineer, University of Siena

Mr Massimo Fabi
Director-General, Parma Local Health Agency

Mr Antonio Ferro
President, Extra Parma Municipality

Ms Carlotta Ferroni
Veterinary Officer, Ministry of Health

Mr Marco Filippeschi
Mayor, Pisa Municipality

Ms Sandra Frateiacci
Federasma

Dr Pina Frazzica
Director-General, Centre for Training and Research in Public Health

Dr Enzo Funari
Chair, Task Force on Surveillance Systems, Istituto Superiore di Sanità

Dr Diana Gagliardi
Researcher, National Institute for Occupational Safety and Prevention

Mr Roberto Garavaglia
Director, Marcegaglia Group

Mr Franco Ghiene

Mr Roberto Ghiretti
Sports Assessor, Parma Municipality

Mr Paolo Giandebiaggi
Architect, Parma Province

Dr Liana Gramaccioni
Administrative Officer, Directorate-General for Prevention, Ministry of Health

Ms Gabriella Guerra
Communication Agency, Rome

Ms Ana Isabel Fernandes Guerreiro
Researcher, Health Promotion Programme, Meyer University Children's Hospital, Florence

Ms Lucia Iannacito
Respiratory Technician, Fondazione Salvatore Maugeri IRCCS

Mr Klaus Ladinser
Environmental Assessor, Bolzano Province

Ms Stefania La Grutta
Prevention and Environment Agency, Sicily Region

Mr Rocco Landi
Administrative Officer, Istituto Poligrafico e Zecca dello Stato

Dr Paolo Laurioia
Head, Regional Environmental Protection Agency

Ms Francesca Lopez
Administrative Officer, Istituto Poligrafico e Zecca dello Stato

Ms Renata Lottici
Oncologist, Parma Province

Mr Pietro Lucchese
Communication Agency

Dr Pierluigi Macini
Head, Public Health Service, Directorate-General for Health and Social Policy, Ministry of Health

Dr Giuseppe Magro
Researcher, Energy, Nuclear Engineering and Environmental Control, University of Bologna

Dr Pietro Malara
Senior Medical Officer, Directorate General for EU and International Relations, Ministry of Health

Dr Stefania Marcheciampani
Biologist, National Institute of Health

Mr Alessandro Marchetti Tricamo
Engineer, Emobility

Mr Paolo Mauri
Director, ASC srl

Mr Gerardo Mauro
Director, Klaus Davi & Co.

Dr Sonia Mele
Technical Officer, Directorate-General for Prevention, Ministry of Health

Dr Maria Teresa Menzano
Medical Officer, Department for Prevention and Communication, Ministry of Health

Mr Arcangelo Merella
Director, Infomobility

Dr Paola Michelozzi
Department of Epidemiology, Local Health Authority, Rome

Dr Antonio Moccaldi
President, National Institute for Occupational Safety and Prevention

Mr Davide Mora
Road Infrastructure Assessor, Parma Municipality

Mr Antonio Moreni

Mr Marzio Flavio Morini
President, Environment Committee, National Association of Italian Municipalities

Mr Antonio Moroni
Italian Ecological Society, Parma Province

Ms Andrea Mozzarelli
Italian Bicycle Federation

Professor Antonio Mutti
Laboratory of Industrial Toxicology, Department of Clinical Medicine, Nephrology and Health Sciences, University of Parma Medical School

Mr Pierantonio Muzzetto
Medical Association

Dr Antonio Navarra
Senior Scientist, National Institute of Geophysics and Vulcanology

Dr Margherita Neri
Head, Pulmonary Rehabilitation Division, Fondazione Salvatore Maugeri IRCCS

Mr Fabrizio Pallini
Health Councillor, Parma Municipality

Mr Francesco Papi
Minister's Press Officer, Ministry of Health

Professor Walter Pasini
Director, WHO Collaborating Centre for Tourist Health and Travel Medicine

Ms Lidia Pavone

Dr Roberta Pirastu
Researcher, Department of Animal and Human Biology, La Sapienza University of Rome

Ms Nicola Pirrone
Director, Institute of Air Pollution

Mr Riccardo Pozzi
Director, Sustainable City, Florence Municipality

Mr Antonio Prade
Mayor, Belluno Municipality

Dr Andrea Ranzi
Project Manager, Prevention and Environment Agency, Emilia-Romagna Region

Mr Alberto Rho
Board Member, Milan Transport Agency

Ms Renata Rizzo
Assistant to Professor Giovanni Berlinguer, University of Rome

Mr Alberto Rochira
Communication Agency

Dr Daniela Rodorigo
Director-General, Directorate-General for Communication, Ministry of Health

Ms Antonia Ronchei
Klaus Davi & Co.

Mr Vincenzo Ruvolo
Prevention and Environment Agency, Sicily Region

Ms Monica Saccani
Parma Municipality

Ms Cristina Sassi
Environment Assessor, Parma Municipality

Ms Francesca Senese
Collaborator, Health and Social Agency, Emilia-Romagna Region

Professor Vittorio Silano
Director-General for Health, Ministry of Health

Mr Carlo Silva
President, Clickutility

Mr Fabrizio Simonelli
Director, WHO Collaborating Centre for Health Promotion and Capacity-Building in Child and Adolescent Health, Meyer University Children's Hospital, Florence

Mr Pietro Somenzi
President, Infomobility

Mr Lorenzo Spizzichino
Technical Officer, Directorate-General for Prevention, Ministry of Health

Mr Alessandro Tassi Carboni
President, Architects Association, Parma Province

Mr Angelo Tedeschi
President, Engineers Association, Parma Province

Professor Stefano Tibaldi
Director-General, Prevention and Environment Agency, Emilia-Romagna Region

Ms Jessica Tuscano

Mr Renzo Valloni
Professor, University of Parma

Mr Sergio Venturi
Director, Hospital Agency, Parma Province

Mr Marco Verdesi
Director, Extra

Dr Roberta Vicentini
Lecturer, Energy, Nuclear Engineering and Environmental Control, University of Bologna

Mr Stefano Zauli Sajani
Prevention and Environment Agency, Emilia-Romagna Region

Luxembourg

Mr Helmut Blöch

Malta

Ms Helen Muscat
Breast Cancer Malta

Netherlands

Ms Regina Aalders
Senior Coordinating Officer, Global Health, Welfare and Sport, Ministry of Health, Welfare and Sport

Mr C.J.M. van den Bogaard
Specialist, Health and Indoor Air Quality, Ministry of Housing, Spatial Planning and the Environment

Ms Nelly van Brederode
Environmental Physician, Centre for Inspection Research, Environment and Health, National Institute for Public Health and the Environment (RIVM)

Professor Bert Brunekreef
Institute for Risk Assessment Sciences, Utrecht University

Ms Sandra van Buggenum
Environmental Health Professional, Zuid Limburg Public Health Service

Ms Maureen Butter
Coordinator, Dutch Platform on Health and Environment

Ms Lisbeth Hall
Researcher, Advisory Service for the Environment and Health Inspectorate, National Institute for Public Health and the Environment (RIVM)

Mr Jeljer Hoekstra
Researcher, National Institute for Public Health and the Environment (RIVM)

Mr Rob Jongeneel
Researcher, Centre for Environmental Health Research, National Institute for Public Health and the Environment (RIVM)

Ms Ellen Koudijs
Researcher, Centre for Environmental Health Research, National Institute for Public Health and the Environment (RIVM)

Ms Hanneke Kruize
Project Manager/Researcher, Centre for Environmental Health Research, National Institute for Public Health and the Environment (RIVM)

Professor F.X. Rolaf van Leeuwen
Centre for Substances and Risk Assessment, National Institute for Public Health and the Environment (RIVM)

Mr Floor Lieshout
Chief Executive Officer, Youth for Road Safety (YOURS)

Ms Merel Linthorst
GGD Nederland

Dr Frank Pierik
Senior Researcher, Environment and Health

Ms Brigit Staatsen
Senior Researcher, Centre for Environmental Health Research, National Institute for Public Health and the Environment (RIVM)

Ms Marjolijn Verschuren
Policy Adviser, Ministry of Housing, Spatial Planning and the Environment

Norway

Ms Carola Bjørklund
Senior Adviser, Ministry of Foreign Affairs

Mr Scott Randall
Research Scientist, Norwegian Institute for Air Research

Ms Aileen Yang
Research Scientist, Norwegian Institute for Air Research

Portugal

Professor Jose M. Calheiros
Deputy Director-General, National Institute of Health

Professor Eduardo Oliveira Fernandes
Joint Research Centre, Institute of Mechanical Engineering, University of Porto Faculty of Engineering

Serbia

Professor Aleksandar Milovanovic
Director, Dr Dragomir Karajovic Institute for Occupational Health, University of Belgrade

Professor Bogoljub Perunicic
Deputy Director, Dr Dragomir Karajovic Institute for Occupational Health, University of Belgrade

Spain

Professor Elisabeth Cardis
Research Professor, Centre for Research in Environmental Epidemiology

Ms Maria José Carroquino Saltó
Senior Researcher, Carlos III Health Institute and WHO Collaborating Centre for the Epidemiology of Environment-Related Diseases

Dr Emmanouil Kogevinas
Centre for Research in Environmental Epidemiology

Professor Mark Nieuwenhuijsen
Research Professor, Centre for Research in Environmental Epidemiology

Sweden

Mr Niklas Johansson
Senior Scientific Adviser, Environmental Assessment Department, Swedish Environmental Protection Agency

Dr Mats E. Nilsson
Senior Researcher, Institute of Environmental Medicine, Karolinska Institute

Switzerland

Dr Emine Nida Besbelli
Consultant

Dr Pierre Studer
Federal Office for Public Health

Tajikistan

Ms Surayyo Saidova
Project Coordinator, Agency for Support of Development Processes "Nau"

The former Yugoslav Republic of Macedonia

Professor Jovanka Karadzinska Bislimovska
Director, Institute of Occupational Health

Professor Vladimir Kendrovski
Head of Sector, Environmental Health, Food Safety and Nutrition, Institute for Health Protection

Ukraine

Professor Yuriy Kundiev
Director, Kyiv Institute for Occupational Health

Ms Svitlana Nigorodova
Adviser, Secretariat of the Minister of the Environment, Ministry of Natural Resources

United Kingdom

Dr Diane Benford
Head, Chemical Risk Assessment Unit, Food Standards Agency

Mr Ben Cave
Director, Ben Cave Associates Ltd., Leeds Innovation Centre

Dr Raquel Duarte-Davidson
Head, International Research and Development Group, Centre for Radiation, Chemicals and Environmental Hazards, Health Protection Agency

Mr Wayne Elliott
Head of Health Forecasting, Met Office

Mr Jon Fairburn
Senior Lecturer, Institute for Environment, Sustainability and Regeneration (IESR), Staffordshire University

Dr Maureen Fordham
Principal Lecturer in Disaster Management, Northumbria University

Mr James Foster
Deputy Chief Inspector (Science and Strategy), Drinking-Water Inspectorate (England and Wales)

Mr John Fintan Hurley
Scientific Director, Institute of Occupational Medicine

Miss Felicity Liggins
Climate Change Consultant, Met Office

Mr Paul Kelly
Department of Public Health, University of Oxford

Dr Anne Matthews
Department of Public Health, University of Oxford

Dr Andy Morse
Reader, School of Environmental Science, University of Liverpool

Professor Virginia Murray
Medical Toxicology Consultant, Chemical Hazards and Poisons Division, Health Protection Agency

Mr Jonathan Peck
Member, National Pest Advisory Panel, Chartered Institute of Environmental Health

Dr Kieron Stanley
Principal Social Scientist, Environment Agency

Professor Stephen Stansfeld
Centre for Psychiatry, Wolfson Institute of Preventive Medicine, Barts and the London School of Medicine and Dentistry, Queen Mary University of London

United States of America

Professor Harvey Brenner
Department of Social and Behavioral Sciences, School of Public Health, University of North Texas Health Science Center

Intergovernmental bodies and international organizations

European Commission

Ms Laura Bellorini
Communications Assistant, Joint Research Centre

European Environment Agency

Mr Jean-Bernard Blatrier

Mr Ove Caspersen
Project Manager, Communication, Corporate Affairs

European Food Safety Authority

Ms Laurence Caratini
Policy Officer

Ms Anna Federica Castoldi
Food Contact Materials, Enzymes and Flavourings Unit

Mr Andrew Cutting
Press Officer

Mr Stefan Fabiansson
Data Collection and Exposure Unit

Ms Anne-Laure Gassin
Communications Director

Ms Kerstin Gross Helmert
Scientific Cooperation Unit

Ms Claudia Heppner
Contaminants Unit

Dr Juliane Kleiner
Head, Nutrition Unit

Ms Rita Lazar

Dr Djien Liem
Scientific Committee and Advisory Forum Unit

Ms Christine Majewski
Strategic Adviser

Mr Stephen Pagani
Executive Director, Head of Press Office

Mr Olivier Ramsayer
Director of Administration

Ms Jane Richardson
Assessment Methodology Unit

Dr Jiri Ruprich
Member, Management Board

Ms Egle Serrao
Administrative Assistant

Ms Claudia Timanti
Administrative Assistant

Ms Luisa Venier

International Commission on Occupational Health

Dr Sergio Iavicoli
Secretary-General

International Federation of Environmental Health

Mr Stephen Cooper
Treasurer

Mr Bernard Forteath
President

Mr Shane Keane
Council Member

International Federation of Red Cross and Red Crescent Societies

Mr Leon Prop
Head of Operations, Europe Zone Office

Ms Sonja Tanevska
Health and Care Coordinator, Europe Zone Office

International Society of Doctors for the Environment

Dr Ernesto Burgio
Scientific Committee Cordinator, Italy

Dr Roberto Romizi
President, Italy

United Nations Development Programme

Ms Katy Norman
Consultant, Human Rights-Based Approach in the Water Sector, Regional Centre for Europe and the Commonwealth of Independent States

World Health Organization

Dr Tahera Emilie van Deventer
Scientist, International Electromagnetic Fields Project

Nongovernmental organizations

Akut

Ms Jean Huss
President

European Child Safety Alliance

Ms Morag MacKay

European Federation of Allergy and Airways Diseases Patients Associations

Mr Giorgio Salerni

European Respiratory Society

Professor Jorrit Gerritsen
Past President

Health and Environment Alliance

Ms Gill Erskine

Ms Anne Stauffer
Policy Manager

Ms Diana Smith
Communications and Media Consultant

Ms Lisette van Vliet
Toxics Policy Advisor

Regional Environmental Center for Central and Eastern Europe

Professor Judit Szaszne Heszlenyi
Head Teacher in Biology and Ecology, Trefort Training School, Eotvos University, Hungary

Ms Tamara Nikolic
Junior Expert, CIVITAS Initiative

Dr Agnes Schroth
Deputy Director, Trefort Training School, Eotvos University, Hungary

Dr Eva Vaskovi
Head, Department for Air Quality Monitoring, National Institute of Environmental Health, Hungary

Women in Europe for a Common Future

Dr Arunas Balsevicius
Director, Station of Nature Research and Environmental Education

Ms Anne Barre
Director, Women in Europe for a Common Future France

Dr Nita Chaudhuri
Environmental Health Promoter/Researcher

Ms Anne-Marie Drieskens
Secretary, Family Policy

Ms Johanna Hausmann
Press and Public Relations Coordinator

Ms Danielle van Kalmthout
Policy Adviser, Confederation of Family Organizations in the European Union

Dr Margriet Mantingh Samwel
Coordinator, Water

Ms Elena Manvelyan
President, Armenian Women for Health and Healthy Environment

Ms Katrina Phillips
Chief Executive, Child Accident Prevention Trust

Ms Olivia Radu
Project Officer

Ms Sara Reekmans
Environmental Health Worker, Logo Limburg

Dr Petr Sharov
Programme Director, Far Eastern Environmental Health Fund

Ms Farida Shorukova

Ms Svitlana Slesarenok
Black Sea Women's Club

Dr Anke Julie Stock

Ms Anna Tsvietkova
Coordinator, Water Issue Group, European Eco-Forum

Mr Umidzhon Ulugov
National Coordinator, Green Patrols Movement, Public Organization Youth of the 21st Century

Mrs Chantal Vandenbossche
Coordinator, Communications

Ms Corinne Zimmer
Scientific Expert

International Youth Network

Mr David Rivett
Network Coordinator

Mr Itziar Badenas Rue
Andorra

Ms Lydia Etzlstorfer
Austria

Mr Dominik Goldnagl
Austria

Mr Stephan Längle
Austria

Ms Roxana Reindl
Austria

Ms Doriane Fuchs
Belgium

Ms Julie Teng
Belgium

Mr Emil Balavac
Bosnia and Herzegovina

Ms Diba Hadziahmetovic
Bosnia and Herzegovina

Ms Desislava Zlatk Taneva
Bulgaria

Ms Nikola Panduric
Croatia

Ms Celie Manuel
Denmark

Mr Malthe Stentoft
Denmark

Ms Kristina Aare
Estonia

Mr Mikhel Raag
Estonia

Mr Tomi Nieminen
Finland

Mr Pascal Conges
France

Mr Martin Rieussec
France

Ms Marika Tsereteli
Georgia

Mr Alexander Karyolaimos
Greece

Ms Evangelia Kontogianni
Greece

Ms Alma Ildikó Almasi
Hungary

Mr Andras Almasi
Hungary

Ms Sorcha Cusack
Ireland

Mr Omri Shaffer
Israel

Ms Silvia Eleonora Gazzani
Italy

Ms Chiara Palieri
Italy

Ms Olga Gallo Stukan
Italy

Ms Asela Muratbeko Ongarbayeva
Kazakhstan

Ms Gulnara Zhenishbekova
Kyrgyzstan

Mr Vytautas Krasnickas
Lithuania

Ms Lara Cassar
Malta

Ms Kristina Miggiani
Malta

Mr Jacob Vella
Malta

Ms Anthia Ann Zammit
Malta

Mr Vladimir Rakocevic
Montenegro

Ms Marijs van Hoek
Netherlands

Ms Michaela Hogenboom
Netherlands

Mr Thijs F. P. Kuijper
Netherlands

Ms Helene Kaltenborn
Norway

Mr Adrian Kowalik
Poland

Ms Agata Dominika Mucha
Poland

Ms Alicja Ewa Naporska
Poland

Mr Bartlomiej M. Tarkowski
Poland

Ms Raquel Sofia Sebastiao Canha
Portugal

Mr Delfim Diogo Ferreira Duarte
Portugal

Mr Tiago Salgaro de Magalhaes Taveira Gomes
Portugal

Mr Gustavo Pizarro Lopes
Portugal

Ms Catarina Marques Ribeiro
Portugal

Ms Laura Nunes Soares Sequeira Salavessa
Portugal

Ms Daria Catalui
Romania

Ms Alina Bezhenar
Russian Federation

Ms Irina Fedorenko
Russian Federation

Ms Evgeniya Soboleva
Russian Federation

Mr Danilo Arsenijevic
Serbia

Ms Jovana Dodos
Serbia

Mr Vulkan Gacaferri
Serbia

Ms Guri Shkodra
Serbia

Mr Blaz Gasparini
Slovenia

Ms Ida Karkiainen
Sweden

Ms Olivia Heller
Switzerland

Ms Marjona Bahraddini
Tajikistan

Ms Martina Karatrajkova
The former Yugoslav Republic of Macedonia

Mr Filip Radevski
The former Yugoslav Republic of Macedonia

Ms Mariana Malashniak
Ukraine

Mr Richard Paul Miner
United Kingdom

Miss Emily-Jane Murrell
United Kingdom

Mr Atong Nyantut William Nyuon
United Kingdom

Mr Kyle Worgan
United Kingdom

Ms Irina Ruslanovn Gilfanova
Uzbekistan

Ms Alexandra Povarich
Uzbekistan

Secretariat

WHO Regional Office for Europe

Mr Roger Aertgeerts
Scientist

Mr Matthias Braubach
Technical Officer, Housing and Health

Ms Pamela Charlton
Editor *(Rapporteur)*

Mr James Creswick
Technical Officer

Ms Lucia Dell'Amura
Administrative Clerk

Ms Tina Charlotte Kiaer
Information Officer

Dr Rokho Kim
Technical Officer

Dr Hilde Kruse
Regional Adviser, Food Safety

Dr Marco Martuzzi
Scientific Officer, Health Impact Assessment

Dr Eva Franziska Matthies
Technical Officer

Ms Geraldine McWeeney
Technical Officer, WHO Country Office, Serbia
(Rapporteur)

Dr Bettina Menne
Medical Officer, Global Change and Health

Mr Francesco Mitis
Technical Officer, Rome

Mr Pierpaolo Mudu
Technical Officer, Rome

Mrs Leda Nemer
Technical Officer

Ms Julia Nowacki
Technical Officer, Rome

Mr Charles Robson
Head, Translation and Editorial *(Rapporteur)*

Ms Cristiana Salvi
Technical Officer

Dr Dinesh Sethi
Technical Officer

Dr Tanja Wolf
Technical Officer, Climate Change and Health